THE THERAPEUTIC COMMUNITY FOR ADDICTS

„I have spent a whole lifetime learning to be pretty careful with people, to be sort of delicate and gentle, and to treat them as if they were like brittle china that would break easily. The first thing that interested me in what was going on here is the evidence that indicates that maybe the whole attitude is wrong. What I have read about Synanon, as well as what I saw last night and this afternoon, suggests that the whole idea of the fragile teacup which might crack or break, the idea that you mustn't say a loud word to anybody because it might traumatize him or hurt him, the idea that people cry easily or crack or commit suicide or go crazy if you shout at them – that maybe these ideas are outdated."

„The process here basically poses the question of what people need universally. It seems to me that there is a fair amount of evidence that the things that people need as basic human beings are few in number. It is not very complicated. They need a feeling of protection and safety, to be taken care of when they are young so that they feel safe. Second, they need a feeling of belongingness, some kind of a family, clan, or group, or something that they feel that they are in and belong to by right. Third they have to have a feeling that people have affection for them, that they are worth being loved. And fourth, they must experience respect and esteem."

„You could say that the kinds of problems we have, the open troubles – not being able to resist alcohol, not being able to resist drugs, not being able to resist crime, not being able to resist anything – that these are due to the lack of these basic psychological gratifications. The question is, does Daytop supply these psychological vitamins? My impression as I wandered around this place this morning is that it does."

„Could it be that Daytop is effective because it provides an environment where these feelings are possible? I have a lot of impressions and thoughts rushing in on me. I've been asking a thousand questions and trying out a thousand ideas, but this all seems to be part of it. Let me say it this way: Do you think that this straight honesty, this bluntness that even sounds cruel at times, provides a basis for safety, affection, and respect? It hurts, it must hurt."

„It seems possible that this brutal honesty, rather than being an insult, implies a kind of respect. You can take it as you find it, as it really is. And this can be a basis for respect and friendship."

„After you get over the pain, eventually self-knowledge is a very nice thing. It feels good to know about something rather than to wonder about, to speculate about it."

„It is good to be able to know."

Abraham H. Maslow (1967)

MARTIEN KOOYMAN

The Therapeutic Community for Addicts

INTIMACY, PARENT INVOLVEMENT, AND TREATMENT SUCCESS

SWETS & ZEITLINGER   PUBLISHERS

AMSTERDAM ▪ LISSE ▪ BERWYN, PA

4

Library of Congress Cataloging-in-Publication Data

(Applied for)

CIP-gegevens Koninklijke Bibliotheek, Den Haag

Kooyman, Martien

The therapeutic community for addicts : intimacy, parent involvement and treatment success /
Martien Kooyman. – Amsterdam [etc]: Swets & Zeitlinger
Oorspr. verschenen als proefschrift Rotterdam, 1992. – Met lit. opg.
ISBN 90-265-1358-5
NUGI 735
Trefw.: verslaving ; hulpverlening / groepstherapie ; druggebruikers.

Printed in the Netherlands by Offsetdrukkerij Kanters B.V., Alblasserdam
Cover design: Rob Molthoff
Photographer: Wouter Thorn Leeson
Sculpture: Peter de Leeuwe and residents of the Emiliehoeve
Cover printed in the Netherlands by Casparie, IJsselstein

ISBN 90 265 1358 5
NUGI 735

Contents

Preface

When I founded the Emiliehoeve Therapeutic Community in The Hague in 1972 I was ignorant and rather naive. I had never seen a recovered addict except at a performance in a theater by residents of a therapeutic community in New York with the name 'Daytop Village'. The play was based on the life-story of an addict. It was called 'The Concept'. At the end of the performance the players descended from the stage and reached out to the audience saying: „Would you love me?" Some years later, I again attended a play perfomed by residents of the Portage program in Montreal. It was at one of the World Conferences of therapeutic communities. The play opened with a girl alone on the stage saying: „I am not perfect, but I am good enough", the same words Dan Casriel used to make people scream in his therapy groups to get rid of messages from the past. Both sentences express in my opinion the essence of the problems behind the symptom of drug addiction: addicts have a negative self-concept and, out of fear of being rejected, do not show their need for love.

In the days when I started the Emiliehoeve program it was hard in those days to find any written documentation on the treatment program in therapeutic communities for addicts. I was thankful for receiving from a colleague a copy of *So Fair a House: The Story of Synanon*, written by the same Daniel Casriel. This book had such an impact on me that, when Casriel visited The Netherlands in 1972 I went to his wokshops. A lot has changed since then. Therapeutic communities for addicts have spread over all continents. Their research publications can compete with those in the mental health field. The outcome studies give a good insight in the success and limitations of the therapeutic community programs. However few efforts have been done to link the processses in the therapeutic communities for addicts with existing theories in the addiction field. In this book I present a theoretical frame of reference in relation to the therapeutic communities, their concepts, treatment tools and techniques.

My process as program director is described in the Chapter on the development of the Emiliehoeve program. The follow-up study of the initial 250 residents of the Emiliehoeve admitted for the first time has been made possible by the contribution of many persons. First of all staff and residents of the 'Emiliehoeve' and the 'Essenlaan' therapeutic communities, the detoxification centers 'De Weg' and 'Heemraadssingel', the day center 'Het Witte Huis' and the induction and re-entry sections of the drug-free programs in The Hague and Rotterdam, staff off the Department of Social Psychiatry and the Institute for Addiction Research of the Erasmus University in Rotterdam and the students who were involved in the interviews.

It was Prof. Cees Trimbos who, when I started the research stressed the importance of focusing not only on treatment outcome, but also describing the development of the Emiliehoeve as the first drug-free therapeutic community founded in the Netherlands. I wish to thank Prof. Joost Schudel for his valuable general suggestions and Prof. Charles Kaplan for his support and supervision of the research part of the study. I greatly appreciated the assistance of Peter Blanken for the statistical analysis of the data and of Gerben van der Heide for the data processing. I wish to thank Marina Smit and Tineke Hanssen for their secretarial work. The study was published as a dissertation at the Erasmus University Rotterdam in September 1992.

April 1993
Martien Kooyman

CHAPTER 1

Introduction.
The therapeutic communities
for addicts

Therapeutic communities for persons with psychiatric problems have developed since the Second World War. It was with the foundation of Daytop Lodge in New York in 1963 (renamed Daytop Village in 1964) that a treatment setting modelled after Synanon, a living and working community of former addicts, also received the name 'therapeutic community' (Sugarman, 1974). Daniel Casriel, the psychiatrist who had been one of the founders of Daytop, identified this form of treatment as a process similar to the therapeutic communities developed by Jones for psychiatric patients, with social learning through social interaction as their main concept (Casriel, 1963; Jones, 1953).

Many treatment programs founded in the United States were modelled after Daytop Village and Synanon. In 1975 the American therapeutic communities for addicts formed an organization: The Therapeutic Communities of America. To be able to establish minimum standards for staff in therapeutic communities, the following list of 10 competencies were described in a manual for therapeutic communities, published by the Therapeutic Communities of America (Kerr, 1986):

1. Understanding and promoting self-help and mutual help.
2. Understanding and practising positive role modelling.
3. Understanding of social learning versus didactic learning.
4. Understanding and promoting the concept of 'no we-they dichotomy'
5. Understanding and promoting upward mobility and the privilege system.
6. Understanding and practising the concept of 'acting as if'.
7. Understanding the relationship between belonging and individuality.
8. Understanding the need for a belief system within the community.
9. Ability to maintain accurate records.
10. Understanding and facilitating group process.

In 1980 at the Fifth World Conference of Therapeutic Communities, held in Noordwijkerhout, the Netherlands, organized by the Therapeutic Community Section of the International Council on Alcohol and Addictions, the World Federation of Therapeutic Communities was founded. One year later the following definition of the therapeutic community was adopted after several years of discussion at international conferences:

> The primary goal of a therapeutic community is to foster personal growth. This is accomplished by changing an individual's life style through a community of concerned people, working together to help themselves and each other.
>
> The therapeutic community represents a highly structured environment with defined boundaries, both moral and ethical. It employs community imposed sanctions and penalties as well as earned advancement of status and privileges as part of the recovery and growth process. Being part of something greater than oneself is an especially important factor in facilitating positive growth.
>
> People in a therapeutic community are members, as in any family setting, not patients, as in an institution. These members play a significant role in managing the therapeutic community and acting as positive role models for others to emulate.
>
> Members and staff act as facilitators, emphasizing personal responsibility for one's own life and for self-improvement. The members are supported by staff as well as being serviced by staff, and there is a sharing of meaningful labor so that there is a true investment in the community, sometimes for the purpose of the survival.
>
> Peer pressure is often the catalyst that converts criticism and personal insight into positive change. High expectations and high commitment from both members and staff support this positive change. Insight into one's problems is gained through group and individual interaction but learning through experience, failing and succeeding and experiencing the consequences is considered to be the most potent influence toward achieving lasting change.
>
> The therapeutic community emphasizes the integration of an individual within this community and the progress is measured within the context of that community against the community's expectations. It is this community along with the individual that accomplishes the process of positive change in the member. The tension created between the individual and this community eventually resolves in favor of the individual, and this transition is taken as an important measure of readiness to move toward integration into the larger society.
>
> Authority is both horizontal and vertical, encouraging the concept of sharing responsibility and supporting the process of participating in

decision making when this is feasible and consistent with the philosophy and objectives of the therapeutic community (Kerr, 1986).

This definition is largely a description of a therapeutic community, modelled after the Synanon Community which was founded in 1958. Although there are similarities with therapeutic communities which had been developed in several European cities to serve different populations of psychiatric patients, there are clear differences. While most therapeutic communities for psychiatric patients, modelled after the unit run by Maxwell Jones in the Henderson Hospital in England, were characterized by a democratic organiz-ation with the concept that every decision concerning the community should involve every individual and should be discussed with everyone present, the American therapeutic community for addicts had a hierarchical staff and residents structure with participation of the residents in decision-making based on the level of their personal growth in the treatment program.

Most therapeutic communities for addicts have a highly structured program. The Synanon Daytop Therapeutic Community has become the model for most therapeutic communities for drug addicts in many countries in all continents. This model has been called the 'concept therapeutic community' (Sugarman, 1974; Piperopoulos, 1977; Schaap, 1980). Referring to the hierarchical structure of organization, the most prominent difference with the therapeutic communities of the Jones model (also referred to as democratic therapeutic communities) is that the concept therapeutic communities are also called 'hierarchical therapeutic communities' (Schaap, 1987). As no substitute drug treatment was given in therapeutic com-munities for drug addicts, they are also commonly referred to as 'drug-free' therapeutic communities, to differentiate them from medical model clinics and methadone programs.

The development of the therapeutic community model, the treatment philosophy, the concepts and the structure of the treatment program, the research literature on therapeutic communities for addicts will be reviewed in this book. The characteristics of families of addicts are described as well as ways of involving parents and other relatives in the treatment program. Starting from the existing main theories of addiction, the treatment process of the therapeutic community is placed in a theoretical framework. Early traumatization, fear of intimacy and rejection and a negative self-concept are linked with the vulnerability for addiction.

The Emiliehoeve, the therapeutic community described in this book, was founded in the Netherlands as a treatment program for drug addicts, initially modelled after the democratic communities for psychiatric (i.e. usually neurotic) patients in the Netherlands following the principles and structure of Maxwell Jones. After the staff of the Emiliehoeve had discovered through experience that this system was not suitable for the addicts admitted, the drug-free hierarchical model of the American therapeutic communities was gradually introduced. In the second year of its existence, the Emiliehoeve Therapeutic Community had fully adopted the structure and tools of the Synanon/Daytop model. However, the philosophy and concepts of this model were introduced after only six months.

An overview of the evolution of the Emiliehoeve program will be given in Chapter 10. From the chaotic first months, the Emiliehoeve developed into a structured program modelled after the American therapeutic communities. Eight distinct phases have been specified and related to the process of treatment. Hierarchial structure and intimate staff-client relations are seen as especially important in differentiating the phases.The follow-up study of residents of the Emiliehoeve program is described in Chapter 11. The outcome of the treatment is related to client and program characteristics. The determinants of success are also analyzed.

Whenever 'he' is used in the text referring to clients or residents it should be read as 'he or she'.

CHAPTER 2

The evolution of
therapeutic communities

During the 1960s Europe was faced with a new problem: a considerable number of young people became involved with the use of illegal drugs. Until then drug abuse, particularly the use of opiates, was almost entirely confined to a small group of people working in the health field such as physicians, nurses and pharmacologists. In most European countries the Opium Law had been operative since the 1920s, long before European society had to deal with the effects of drug addiction. In most European countries, drug addiction was seen as a health problem. Drug addicts were referred to physicians, psychiatrists, mental hospitals and clinics for alcoholics (Kooyman, 1984a). In the mid 1960s the drugs most commonly used by youngsters were cannabis, LSD, amphetamines and opium. In most European countries heroin was not introduced before the 1970s. In Great Britain heroin was prescribed to adolescent drug addicts by physicians and from 1968, by special clinics linked with psychiatric hospitals or psychiatric wards of general hospitals. Actually, this so-called British system never was a system; it was a result of the British tradition to allow doctors to prescribe drugs for their addicted patients. Despite the availability of opiates through physicians, heroin began to be sold on the British black market in 1971 (Edwards, 1979).

In the Netherlands, where in 1968, methadone was introduced in maintenance programs for youngsters addicted to opium, heroin appeared on the black market in 1972 and soon replaced amphetamines and opium to become the most popular hard drug. As heroin and methadone programs did not cure patients of their addiction, and the traditional general and psychiatric hospitals proved to be even less successful in this respect, new methods to treat the growing addict population had to be developed.

The origin of the American therapeutic communities

The roots of the American therapeutic communities originated in Synanon. This self-help community was influenced greatly by the Alcoholics Anonymous (AA) movement. The founders of AA had strongly been involved with the Oxford Group, a spiritual organization which after 1938 became known as the Moral Rearmament Movement. This movement had a religious character. It proclaimed, amongst other things, self-examination in group meetings (Glaser, 1977).

The founder of Synanon was Charles Dederich, a former businessman and a former member of AA. He changed the AA meetings, which took place in his home, into groups in which members confronted each other in a direct, emotional way. He called these groups 'games'. The regular participants managed to abstain from alcohol and drugs. They decided to start a community where they could live together. In 1958 they founded this community in Santa Monica, California and gave it the name Synanon.

The games became the heart of the community. The acceptance of the expression of verbal aggression between group members as a therapeutic element of the games could most probably develop because of the absence of professionals. The concept of self-reliance described by the American philosopher R.W. Emerson greatly influenced Dederich's ideas. His vision was to create a better society. The residents of the Synanon community did not want to return to the society which they denounced. Within Synanon they created an ideal society. The residents had no private property, worked for seven days followed by seven days free time and adopted several other patterns. At any time, work could be stopped to have a 'game' to resolve conflicts. There was no money in the community. They ran several businesses and farms. Synanon became a fascinating social experiment. However, the experiment itself ultimately failed, due to a lack of sufficient control by the leader. Among other things, Dederich ordered couples to change partners and to get sterilized and finally allowed physical violence in dealing with juvenile delinquents who were referred to their community. Synanon collapsed but it gave birth to the stormy development of the therapeutic communities for addicts (Deitch, et al., 1979).

Ex-residents from Synanon established therapeutic communities which, unlike Synanon set resident's return to society as the goal. Some of those self-help programs such as, for instance, Delancey Street in San Francisco managed to survive without any financial support from the government. However, more widespread was the application of the model by government sponsored organizations. The psychiatrist Daniel Casriel and the criminologist Alexander Bassin followed the suggestions of J.L. Moreno, who developed psychodrama as a method of group psychotherapy (1959), of Carl Rogers, who developed his 'Rogerian' psychotherapy and was later leading encounter groups (1970), and of the sociologist and psychodramatist Lewis Yablonsky (1967), to pay a visit to Synanon. Their ultimate aim was to set up a treatment center for the growing number of addicts in New York. Casriel decided to stay in this community for several days. Synanon housed at that time more than one hundred ex-addicts, the majority of them being ex-heroin addicts who apparently were successfully living together without relapsing. Life in this community had little to do with the friendly, understanding approach practised in most professional treatment centers of those days. In the early 1960s this was a unique situation (Casriel, 1963).

Rogers had pointed out to Casriel that criticism of others and emotional reactions to one's behavior by others in the group were most probably seen by the residents as care and concern.

After Casriel's return to New York, Daytop Village was founded in 1963. Contrary to Synanon, the goal was to restructure the residents' life in such a way that they could function again in society without a need for drugs. Daytop Village, and Phoenix House founded some years later, were models for a quickly growing number of therapeutic communities in North America (Broekaert, 1976). Outside the United States in Montreal, the Portage program was founded by staff from Daytop Village with John Devlin as the first director. With the help of Daytop Village Bob Garon founded Dare in Manila Phillipines, a therapeutic community not dependent on government funds. Aso in the United States self- supporting therapeutic were established by ex-Synanon members such as Delancey Street in San Francisco and Habitat in Hawaii. Ex-Synanon members were impotant leaders in the early years of the american therapeutic communities, such as Deitch and Brancato in Daytop and Natale in Phoenix House. Professionals ,such as Rosenthal and De Leon in Phoenix House, Densen Gerber in Odyssey House, Biase in Daytop Village, Sholl and Holland in Gateway House, Ottenberg in Eagleville Hospital and Vamos in Portage contribted greatly to the scientific evaluation and development. Under the leadership of Msgr. O'Brien, who was the president of the World Federation of Therapeutic Communities for more than a decade Daytop Village helped to establish many therapeutic communities in Europe and South East Asia.

The origins of the therapeutic communities in Europe

Professionals from Europe visited the United States. They saw the American therapeutic communities and returned with the idea to set up similar centers in their own countries. One of them, Ian Christie, founded Alpha House in 1970 in Portsmouth, England. Some months later Phoenix House, London was founded by Griffith Edwards. An American ex-addict and graduate of Phoenix House, New York became its first director. In 1972, 'Emiliehoeve' was founded in The Hague in The Netherlands. During the first months the program followed the concepts of the therapeutic communities that had been developed within the psychiatric hospital model of the Henderson Clinic, in which Maxwell Jones had departed from the traditional doctor/patient relationship used within psychiatry. Although there was some knowledge about the American self-help therapeutic communities, those models were seen as too authoritarian and too rigid to be applied in the liberal society of The Netherlands (Kooyman, 1975a).

Maxwell Jones' idea to free the patient from the pathogenic, regressive atmosphere of the mental hospital was not new. As early as in 1800, Pinel introduced his 'no restraint' therapy in France. This treatment was based upon no violence, work therapy, and the provision of recreational facilities, etc. Conolly started a similar system in England in 1850. In 1895, open institutions for juvenile delinquents were also founded in England. In Germany, Herman Simon, who was inspired by Bleuler in Zurich, developed activation therapy, teaching patients to accept responsibilities. Most of these innovations were, after some time, replaced by a traditional authoritarian system, especially under pressure of larger numbers of patients and lack of staff (Kooyman, 1978b). Sullivan had stated in 1937 that a mental hospital could become a school for personality growth rather than a custodian for personality failures (Sullivan, 1937). Despite this realistic idealism, the psychiatric institutions remained opposed to radical changes.

During the Second World War, in England, several psychiatrists worked in the Army Selection Unit where they were confronted with soldiers returning from the battlefields suffering from mental breakdown. These men were sent to the Northfield Hospital where each ward was run by a psychiatrist more or less in his own way. One of them, Maxwell Jones, stressed the importance of the participation of the patients in decision-making. He called this method of working 'democratic therapy'. The official goal of the treatment was to cure the patients so they could return to the battlefield. The doctors, however, sent their patients home instead.

One of the psychiatrists, Bion, was asked to introduce group-therapy with the purpose of rehabilitating the patients so they could to return to war. Meeting excessive resistance, this project had to be stopped after six weeks. The authorities then asked Harold Bridger, an army officer and former teacher, to take over the project. He accepted on the condition that he would be responsible for the management of the entire hospital. His first action was to empty one ward, call it 'the club' and have his office next to the main room. Bridger waited in his office suggesting to the patients that they should create a social club in the empty space. After some weeks a social meeting place for all patients of the hospital developed (Bridger, 1985). One of the psychiatrists in the hospital, Tom Main, called it a therapeutic community (Clark, 1985).

Sometime afterward, Jones turned a unit of the Henderson Hospital into a therapeutic community with daily meetings of staff and patients. In what became the new Maxwell Jones model, the patient was given an important role in the therapeutic process. The roles of staff members as well as of patients were now open for discussion. This took place in daily community meetings of staff and patients. Jones made five basic assumptions for therapeutic communities (Jones, 1953):

1. a two-way communication at all levels
2. decision-making at all levels
3. shared leadership
4. consensus in decision-making
5. social learning by social interaction, here and now

The environment was considered to be sufficiently therapeutic so that no individual treatment plan was made. With only one treatment format, the program became selective, leaving the majority of the psychiatric patients in the regression-producing environment of the traditional hospital.

Traditional hospitals cause patients to feel isolated, cast out and hopeless. A major purpose of the therapeutic community model was to prevent this by giving people a feeling of self-esteem as a result of their accepting responsibility. However, contrary to the American therapeutic communities, the Maxwell Jones model treats the individual as a patient up to his discharge without giving him the opportunity to become a staff member himself, if he so wishes.

The Emiliehoeve therapeutic community and its roots

The Emilehoeve therapeutic community is an example of a program founded by professionals that initially applied the principles described by Maxwell Jones and later changed into a hierachically structured program using the concepts of the american terapeutic communities. In the early days of the Emiliehoeve Therapeutic Community it was discovered that the democratic principles of the Jones' model, when applied in the way it had developed in the therapeutic communities for psychiatric, mainly neurotic, patients could become anti-therapeutic. Patients stayed in a regressed state if staff did not apply enough pressure towards making them act responsibly. Staff was surrendering power instead of delegating it, and as the group as a whole was made responsible for everything, individuals avoided their own responsibilities.

During the first months of its existence, decisions at the Emiliehoeve therapeutic community were made by staff and residents together in consensus or by a one-man-one-vote system. In making up the plans for the day, the voting usually resulted in going to the beach, to a coffeeshop, or staying in bed, but generally not in going to work or having group therapy. Since this behaviour was not challenged in this democratic process, conflicts were avoided. The Emiliehoeve staff had not taken into account the level of growth of the residents at that time. As problems were discussed and group sessions were run along psycho-analytical lines, emotions were not dealt with in the group. This resulted in outbursts of violence or other acting-out behavior.

After those chaotic months, the staff decided to take a closer look at the American programs. Two staff members went to an encounter marathon workshop, run by the first director of Phoenix House, London who had left Phoenix House and had become involved as a therapist in the Human Potential Movement. Encounter groups had been developed from the Synanon games and had become the main therapeutic tool of the American therapeutic communities. Impressed, the two staff members introduced encounter groups in the Emiliehoeve Therapeutic Community immediately after their return from the workshop. At the same time, the importance of the concepts of honesty and openness were emphasized. From that moment on, guilt feelings were treated as a normal mechanism; something that should be talked about openly to get necessary relief, instead of being psychiatrically interpreted as a potential symptom of depression.

Gradually, with the help of staff and ex-staff members of Daytop Village and Phoenix House, a clear and structured program was developed. The staff overcame the initial resistance towards creating residents' departments with a hierarchical structure. The staff became aware of the fact that it was important to create conflicts that could be worked out emotionally in the encounter groups where no ranks existed. New residents were now usually seen as unreliable, dishonest, manipulative, egocentric and emotionally immature. They noticed that it was of vital importance for the character-disordered resident with his pseudo-adult image, that the maturation of this 'child' should take place in a structured therapeutic environment, . No longer were residents seen as being on a equal level regardless of the time they had spent in treatment.

The development of therapeutic communities for addicts in Europe

Alpha House and Phoenix House were established in England in 1970. The Emiliehoeve therapeutic community was the first drug free therapeutic therapeutic community founded on the continent of Europe in 1972. With the help of the Emiliehoeve staff, several other therapeutic communities were established in The Netherlands. Among them Breegweestee near Groningen, which in its turn initiated the transformation of the neighboring clinic for alcoholics (Hoog-Hullen) into a structured therapeutic community (Schaap, 1978). Other therapeutic communities for addicts were established in The

Netherlands (Parkweg i.e.), as well as in Sweden (Vallmotorp); they had either started from medical clinic models or from the Maxwell Jones' model. Step by step, these programs became more structured, since the staff realized that it was important for the residents to learn to deal with limits.

In the beginning 'the staff of the European therapeutic communities consisted almost exclusively of professionals. One reason was that there were no ex-addicts available. Another reason was that it took time to convince boards and authorities of the fact that recovered addicts could become reliable staff members. European professionals went to the United States and Canada to learn from the experience of the North American therapeutic communities which are based on the Synanon self-help principles and known as concept therapeutic communities. Europeans learned the importance of using older residents as role models and of introducing encounter and confrontation groups into the program.

Therapeutic communities developing in Western Europe in the mid 70s sent staff to already existing therapeutic communities. With the help of the Emiliehoeve therapeutic community in The Hague and the Essenlaan therapeutic community in Rotterdam, the therapeutic communities De Sleutel and De Kiem in Belgium were founded. The Emiliehoeve therapeutic community was instrumental in helping to develop. many other therapeutic communities in Europe. Staff from a therapeutic community in Bern, Switzerland called Aebi Hus, visited Emiliehoeve, learned about Synanon and after their visit discovered some weeks later that Dr. Karl Deisler – a pediatrician once living near Synanon and a doctor there for some time – was living some miles from their community. Since then he became involved in restructuring their program.

Staff from an ambulatory program in Rome, the Centro Italiano di Solidarietà (C.E.I.S.) visited the Emiliehoeve program in 1976. They were brought in contact with Daytop Village and started their fast-growing movement of therapeutic communities in Italy by organizing the Third World Conference of Therapeutic Communities in 1978, a few months before they opened their first therapeutic community. The CEIS program in Rome helped to establish many therapeutic communities in most regions in Italy and also in Spain. Their training institute in Rome offerred basic and specialized courses for staff from Italy and many other countries within and outside Europe.

Programs sent staff to be trained at the Emiliehoeve therapeutic community. Staff were sent not only from other Dutch programs, but also from programs in Belgium, Sweden, Germany, England, Austria and at the beginning of the 1980s from Greece. Most of these therapeutic communities are more or less structured according to the Daytop and Phoenix House model. In Norway therapeutic communities developed from psychiatric hospitals. In 1982 a concept therapeutic community was set up with help of

Phoenix House, London. That program also helped to start a Phoenix House in Germany. Most therapeutic communities in Germany however, developed rather independently. Even Daytop Germany, founded by Osterhues one year after the Emiliehoeve therapeutic community had been established, had at its inception, not much more in common with Daytop Village in New York than its name. In Berlin several forms of therapeutic communities were developed by social workers. Other German as well as French-speaking therapeutic communities were established in Switzerland.

In France and Denmark structured therapeutic communities hardly developed, or only had a short existence. In France the American therapeutic community model was considered not to comply with the democratic principles of most workers in the field. Because of the fact that the French authorities did not support the development of therapeutic communities, communities for drug addicts of a more or less sectarian character were able to develop, such as the Patriarch communities, which also spread into Spain and the south of Belgium. In Italy rural therapeutic communities developed with emphasis on work and social control. In Italy and Spain therapeutic communities were established with exclusively professionals in the staff as opposed to therapeutic communities wihth mixed staff of professionals and ex-addicts.

Many therapeutic communities applying the hierarchical structure of the American therapeutic communities were faced with strong opposition from professionals as well as politicians. World War II had made people suspicious of the misuse of hierarchical structures and critics of the therapeutic communities were afraid of the potential abuse of power in the system. There is indeed, a real risk of abuse of power in a hierarchical structured therapeutic community. It is important to create control systems reducing this possibility. In the programs, it is important to show that a structure in a therapeutic community is a tool and not an ideology, and that residents are also trained to cope with democratic decision-making in the last phase of the program, the re-entry phase.

Almost all therapeutic communities described above have been in contact with each other and were mutually influenced. Separate from this movement, therapeutic communities according to the Makarenko model based on a socialistic ideology which considers working together as the main therapeutic element, were developed in Eastern Europe and in Germany; those therapeutic communities did not call their model 'therapy' (Broekaert, 1981a).

In 1977 a treatment facility for drug addicts was set up in Poland within a psychiatric hospital; the therapeutic community, as part of this facility was called 'Synanon'. This name was given after a Canadian therapist had visited the program. Although the group sessions were similar to encounter groups, the staff had never heard of Dederich. From 1978, Monar Centers, self-help agricultural therapeutic communities, were established in various places in

Poland. Their structure showed remarkable resemblance with Synanon in the United States. In Yugoslavia and Czechoslovakia professionals established therapeutic communities for alcoholics with a mutual help ideology.

The conferences of the World Federation of Therapeutic Communities are important occasions to meet and to exchange ideas and experiences. The first European Conference of Therapeutic Communities was held in 1982 in Eskilstuna, Sweden. The main theme of the Eskilstuna conference was the third generation of therapeutic communities. The second generation of therapeutic communities developed in Europe copying the principles of the first generation, of the Daytop and Phoenix House programs. Third generation therapeutic communities were seen as programs which added new therapeutic elements to the original concepts.

The professionals of the European therapeutic communities introduced group therapy developed from the same belief in the possibilities of human beings to grow and change that is basic to the philosophy of the North American therapeutic communities. Psychodrama, Trans-actional Analysis, Primal Scream, Bio-energetica and New Identity Process or Bonding groups were added to the programs. Other developments were the introduction of family therapy and individual psychotherapy for residents in the re-entry phase of the program. Therapeutic communities initiated varying structures enabling different populations to benefit from a therapeutic community program. As each therapeutic community offers just one type of treatment, the model cannot be applied for all kinds of people. The challenge of the future is to discover which type of therapeutic community model can best be matched to each individual.

While most therapeutic communities for addicts outside Europe are highly structured with hierarchical staff and residents structures, many European therapeutic communities use a more egalitarian model. The latter are more like the therapeutic communities in the psychiatric field, having a democratic structure (Jones, 1979 and 1986, Zimmer-Höfler, 1981). Therapeutic communities of over twenty-five residents have usually adopted the hierarchical structure of the American programs. European therapeutic communities are in turn, influencing North American therapeutic communities by, for instance, pointing out the importance of introducing creativity into the program. Also, they show that residents can learn skills, such as gardening, farming and printing during their stay in the program. In Europe, as well as in the United States, detoxification centers linked to a therapeutic community were established. Also day centers and evening programs were founded based on the drug-free therapeutic community concept. Therapeutic communities in Europe have started to expand into fields of self-destructive behaviors other than drug addiction. The concepts of a hierarchical therapeutic community could be successfully used in the treatment of alcoholics (Schaap, 1987).

CHAPTER 3

Treatment theory and philosophy

The etiology of addiction

One of the problems in the treatment of addiction is that one can look at the problem from a great variety of viewpoints: as being mainly a problem of biochemical dysfunctioning of the body, a problem of anti-social or deviant behavior of a person, as a result of inborn or acquired vulnerability to addiction, as a result of a pathological family interaction, or as an illness. Trimbos (1980) pointed out that the many definitions of addiction usually say more about the viewpoint of the person who made the definition than about the phenomenon itself.

Most theories on addiction are concerned with the etiology. Few describe addiction as a system. By describing a possible cause we do not get a definition of addiction. It is like describing the condition 'fever' by explaining how a person can get a fever by being infected by a virus or bitten by a malaria mosquito, by having pneumonia or an infection of the urinary tract, or by suffering from a heat stroke. It does not tell us much about the fever itself. It tells us only about the many ways in which we can get a fever. Addiction, in the same way as a fever, is a condition caused by many factors which can interact and reinforce each other. Addiction is not a disease just as fever is not a disease, but it is rather a symptom of an underlying disturbance. The nature of this disturbance can be psychological, interpersonal and/or social. Psychological problems can vary from a character disorder to neurotic or psychotic disease.

Addiction can be considered as a self-administered medication to diminish feelings of tension or pain from a large variety of origins. There may exist an underlying psychological disorder, such as a schizophrenic or affective psychosis, a borderline syndrome, and so on (Kooyman, 1986a). In the case of such a severe underlying psychiatric disorder, successful treatment in a regular therapeutic community for heroin addicts cannot be expected. The behavior of such patients is too complicated for the staff and fellow-residents to deal with, so their admission usually results in rejection and expulsion from the therapeutic community (Lakoff, 1978). Most residents of therapeutic communities belong to the character-disorder type of heroin addicts. The most prevalent lifetime disorder, other than substance dependence, was found in a research in the detoxification center linked with the Emiliehoeve therapeutic community, to be the antisocial personality (Hendriks, 1990).

The many theories on the etiology of addiction can be divided into biological, psychoanalytical, behavioristic, system-oriented, social and self-medication theories.

Biological theories

The biological theories focus on a possible existence of inborn or acquired physiological conditions making a person vulnerable for an addiction. The discovery of a hereditary component in alcohol addiction supports this theory. It is part of the philosophy of the AA, advocating to its members life-long abstinence (Mullan, 1989). Legalization of an addiction to methadone was based on a biological theory, the metabolic deficiency theory of Dole and Nyswander. They based their theory on the fact that they found in their methadone maintenance program that when craving for heroin was eliminated by methadone, the antisocial acts previously exhibited by their clients disappeared (Dole, Nyswander and Warner, 1968). Change of the life-patterns of heroin addicts treated in methadone maintenance programs were evaluated later and the potential negative side-effects of this programs led the inventors of the method ten years later to the following conclusion:

> „Perhaps the limitations of medical treatment for complex medical-social problems were not sufficiently stressed. No medicine can rehabilitate persons. Methadone maintenance makes a first step possible towards social rehabilitation by stabilizing the pharmacological condition of addicts who have been living as criminals on the fringe of society. But to succeed in bringing disadvantaged addicts to a productive way of life, a treatment program must enable its patients to feel pride and hope, in accepting responsibility" (Dole & Nyswander, 1976; Dole, 1988).

If addiction was caused only by biological factors, compulsory detoxification could be the answer. The truth is that treatment really starts after detoxification has taken place and understanding how to prevent a relapse is more important in the treatment of addiction than how to detoxify the addict.

The discovery of endorphines, morphine-like substances in the human body, led to speculations that disturbances in the endorphine system could cause a need to consume endorphine-like substances when available. Endorphines may play a role in the etiology and maintenance of addiction behavior. These endorphines can stimulate self-injecting behavior whereby the substance is injected in the cerebral fluid. Endorphines can be regarded as physiological tranquillizers (van Ree and Fraenkel, 1987). Opiates can replace endorphines on the receptors in the body. When the opiate is no longer administered to the opiate receptors, the morphine is not yet replaced by the endorphines that have not been produced sufficiently by the body, leading to abstinence symptoms (Noach, 1980).

Apart from the discovery of the existence of morphine-like substances produced in the body, animal behavioral experiments provide interesting results that help to understand both the etiology and the process of addiction and other deviant behavior in man. Animals in laboratory experiments can become addicted to morphine. Rats preferred the drug to food (Wecks, 1962). It could be well established that animals do inject themselves with the same drugs that humans abuse and not with drugs that humans do not abuse (Griffith et al., 1980). Hadaway found that rats in a more natural environment, instead of the small box used in laboratory experiments, did not get addicted to morphine when exposed to the substance (Hadaway, 1986).

Rhesus monkeys that had been separated from the mother after birth and were kept socially isolated, as well as monkeys that had been separated at different intervals from the mother, were found to consume more alcohol when exposed to this substance than the non-isolated or separated controls (Van der Kolk, 1987). Van der Kolk cites research by McKinney who found extreme violent reaction to the administration of alcohol to isolated monkeys that was not present among normal controls. Separation from the mother causes both in animals and in human beings, a panic reaction. It is apparently a painful experience. In laboratory experiments, the separation cry of animals disappears after administration of a certain dosage of morphine which does not affect the general behavior. After administration of the opiate antagonist naloxone, the separation cry returns. Other substances such as valium, barbiturates, amphetamines, alcohol and meprobariate were not able to repress the separation cry. Apart from morphine, imipramine and clonidine also had a similar effect. It is worth mentioning here that clonidine is effective in reducing the abstinence symptoms of opiates. When monkeys were given naloxone this increased their attachment behavior. Van der Kolk (1987) also mentioned that social isolation of animals was found to increase the number and activity of the opiate (endorphine) receptors in the brain. The areas with the highest density of opiate-receptors are those which in laboratory experiments, were found to be important in sustaining social relationships. The brain circuits involved in separation are those related to the experience of pain.

Experiments with rhesus monkeys have shown that young animals preferred the fur of an artificial mother to an iron mother with a milk bottle (Suomi and Harlow, 1970; Harlow, 1971). Bowlby, among others, has reported that children in the same way as young monkeys showed an increased tendency to cling to the mother when they experience the threat of danger. This, as in children, persisted even when the mother was brutal, neglectful and representing a danger herself (Bowlby, 1984). When monkeys were only isolated during the first few months of their life and were then brought into contact again, they developed normally. If this was not done or was done later in life, they showed serious abnormal behavior (Griffin and Harlow, 1966). The abnormal behavior was characterized by aggressive behavior to other monkeys and also to their own offspring. They even tried to kill their babies and they were incapable of normal sexual behavior (Harlow, 1971).

Besides this, Suomi (1984) found a critical age as well in relation to both the acute despair reaction and the long term effects. He found that monkeys separated 90 days after their birth showed more abnormal behavior than those separated after 30 or 120 days. This seems to be a critical period in the development of monkeys, in which they are more vulnerable to the effects of separation and subsequent isolation. If the monkeys had been separated from their mother, but were kept together with their peers, they showed less despair behavior. They also behaved normally after 3 to 4 years. Only in stress situations did they behave abnormally again (Suomi and Harlow, 1972). Mason (1967) found in studies with chimpanzees that the young ones played when the mother was near. On being separated, they did not play and after being placed back with their mother, they clung to her; the longer they had been separated, the more they clung to the mother. After amphetamines had been administered to the young ones – making them physiologically more alert – the playing disappeared and the clinging increased. Mason postulated that the physical contact with the mother probably stimulated the opiate system in the body, thus neutralizing the effect of the amphetamines. In his opinion, play stimulated the noradrenergic system.

The reason for referring rather extensively to this animal research is that it provides evidence that, in animals, due to traumatically painful experiences, a variety of behavioral abnormalities occurs characterized by anti-social or violent behavior and inability to take care of their offspring after birth, following artificial insemination. Drugs with abuse liabilities as well as neurotransmitters play a role in these behavior patterns. There seems to be a critical period which could coincide with the myelination of those areas of the brain related to attachment behavior. Not only a psychological traumatization can occur but also a physiological traumatization that may lead to receptors becoming more numerous and hyperactive (Kraemer, 1985). When traumatized monkeys are mixed with normal peers they recover from their obvious abnormal behavior. We can speak of peer treatment and social learning by social interaction when we look at the analogy of treating drug addicts in therapeutic communities.

Biological theories and research may offer us more insight into the etiology of addiction. In people who become addicted, a vulnerability which can be biologically explained, may exist through early traumatization, which results in abnormal behaviors, one of them being substance abuse, as an attempt to overcome the handicap.

Psychodynamic theories

Early psychoanalytic literature focused on the perceived relationship between drug abuse and libido. Freud considered the use of narcotics a substitute for sexual satisfaction; his followers saw addiction as a result of a disturbed childhood. Parts of the personality had not developed normally and got stuck in the 'oral' phase, in the first years of life when the mouth was the most important organ and the mother figure was especially important as the provider of food (Platt, 1986). Analogical to being fixated at the oral phase, persons who inject drugs could be seen as regressed to an even earlier phase in their development, the umbilical phase.

The dependence on the effect of heroin and the involvement of the accompanying lifestyle can be regarded as an adaptation to an impaired ego-development. A heroin addict of the character-disorder type suffers from defects in the pre-genital development of the personality. But, particularly in the separation/individuation stage, the period in which feelings of self-esteem, control of impulses and the super-ego function are developed, a critical period can be found in the family situation. It is usually in this period that the child has received inconsistent messages and that unclear limits were set to his behavior; the parents were frequently absent and often there was a serious emotional deprivation. In this situation the child develops guilt feelings and thinks that he has caused the loss of affection. As a result, painful effects are linked to a feeling of hopelessness, and of being worthless and unlovable. Often an extreme separation anxiety develops in the absence of an internalization of positive object relations and an impaired super-ego development is unavoidable (Hollidge, 1980). The foundation is then set for a serious inhibition about separation from the parents' home in late adolescence. Use of drugs can meet the needs to control this critical phase in life in a narcistic way. The need for affection and support from a parent can be met immediately. Accompanying feelings of guilt are linked with the notion of not being entitled to good feelings and affection. The unavoidable misery linked to drug use is a symbolic solution for this internal conflict. The addict is as dependent on the positive experience of the drug use as on the accompanying misery (Kooyman, 1985b). Prior to their first drug use, these addicts display a behavior which serves as a protection against unconscious, painful feelings. The subsequent use of drugs helps them to attain the same goal.

The addict is unable – and already was so before his drug use – to ask other people for help in a direct way. He controls the situation through manipulative behavior, protecting himself against rejection. His behavior is aimed at immediate gratification. Addicts have no trust in the future; they have a basic distrust of others as well as an absolute lack of self-confidence. Contrary to common belief, addicts do not enjoy the pleasures of life. They fear success and tend to behave in a destructive manner. This fear of success in fact is based on the thought construction that success cannot last very long, that it will be taken away. To eliminate this fear the addict tends to destroy his success. The addict can be regarded as a person who is unconsciously trying to be an infant again, completely dependent, without responsibilities. Regression to a narcissistic state, characterized by immediate gratification of primary needs, is the result of the early childhood experiences in which individuation and separation could not develop in an atmosphere of trust and basic security (Khantzian, 1980).

Szasz (1958) offered an explanation for addiction as the result of a counter-phobic mechanism by which the addict seems to dramatically and repetitively re-enact situations of danger in which the ego hopes to achieve mastery. He also suggested the importance of the super ego which aims at ritually persecuting the drug user (Szasz, 1975). Wurmser argued that drug addiction differs from occasional drug use and can be seen as a symptom embedded in a overall life-style (Wurmser, 1972).

In research conducted by Kaplan and Wogan (1978) using the psychoanalytic concepts, a unitary type of addictive personality could not be found applicable to heroin addicts. According to the authors, the wide variation in addict character-types makes the concept of an 'addict personality' according to the authors probably counter-productive. Treatment appropriate for one character-type may be disastrous for another. In their research the super-ego factor, related to the family structure, accounted for most of the variation in the addiction of heroin users. The role of id and ego factors seemed secondary to that of the super-ego.

Even in 1933, Rado considered a high level of tension and little tolerance to pain to be essential for addicts (Rado, 1933). Casriel, who was his pupil at the Columbia University's Psychoanalytic Clinic for Training and Research (where the Adaptional Psychodynamic Theory with its concept of the fight-fight response to a threatening situation was developed) considered the addict as pain-dependent. Rather than reacting in a fight or flight response the addict reacts by emotionally freezing and remaining in an encapsulated position as a response to danger. In 1962, Casriel was one of the founders of Daytop Lodge, the first therapeutic community for addicts in the United

States sponsored by Government Funds. Using his experience at Daytop, Casriel developed a new kind of psychotherapy, called bonding psychotherapy or the New Identity Process (Casriel, 1972). The Adaptional Psychodynamic Theory had departed form Freud's concepts of inborn libidinous drives, being basically of sexuality and aggression. Aggression was not seen as an instinctive drive but rather as a defense against danger, real or imagined. Casriel defined bonding as being emotionally and physically close to another person. He considered bonding as a primary biological need, such as food, oxygen and sleep. Fulfilment of these needs gives pleasure. Lack of fulfilment results in pain. Also thoughts of fulfilment can arouse pleasure and thoughts of lack of fulfilment can arouse pain. Lack of bonding makes a person vulnerable to a great variety of psychopathology, one of these is addiction. The therapy developed by Casriel is far different from the traditional psychoanalytic therapy. He stressed the importance of showing emotional involvement by the therapist and advocates self-disclosure while in traditional psychoanalysis the therapist is detached, benign and objective. Bonding psychotherapy has become a useful treatment method and has been successfully introduced in therapeutic community programs.

Traditional psychoanalysis however has been largely unsuccessful in the treatment of addicts. Krystal (1970,1988) describes drug addicts as early traumatized persons. Their transference to the therapists is that of an idealized omnipotent mother. They usually react to painful clarifications of the therapist by using drugs again. They are not enough in contact with their emotions to make traditional psycho-analysis, using transference phenomena in the therapeutic relation, possible. Bratter (1973) recognized the need for psychotherapists to become personally involved with addicted individuals in order to be successful in their therapy. Glasser (1965) mentions that without such an involvement there can be no basis for any therapeutic relationship. Khantzian (1982) noted that a major appeal to opiates is their anti-aggressive action. He suggested that addicts who had been exposed to physical abuse and violence in childhood, possessed strong feelings of aggression and sadism towards others. Using opiates made it possible to escape from the dysphoria associated with anger and rage allowing them to remain calm and relaxed.

One of the psychodynamic views on addiction is the self-esteem theory. In this theory developed from Adler's individual psychology, self-esteem is seen as the major psychodynamic mechanism underlying all drug use and abuse. It postulates that all behavior is medicated by the individual's attempt to protect the 'self' within the social environment. Self-esteem develops slowly during the socialization process. The foundation is developed early in life and is present at the time the prototype of the personality is formed. This does not mean that self-esteem cannot be changed positively or negatively later, since the individual is very much responsive to social pressure. Self-esteem develops through exponential behavior involved in mastery of situations and achievement of personal goals. Low self-esteem may result from setting goals too high or from not achieving realistic goals, because there is a lack of confidence in the ability to attain them. The latter may happen when a parent or significant other person does everything for the child, never allowing him of her to develop talents for mastery (Steffenhagen, 1980).

Low self-esteem may result from an overprotective mother, taking away all responsibility for the child's behavior. Approval and rewards are seen as coming only from superior performance, which leads to the setting of high goals and to the development of compensatory safeguards, such as, „If I had not had a headache, been drunk, or used drugs, I would have done better in my exam." Low self-esteem may also develop in a situation where a child gets no attention from the parents as may be the case in an impoverished environment with neglectful parents. A person who has developed low self-esteem feels a need to protect this poor self-image, through compensatory mechanisms which creates further problems in interpersonal relations, adding to the feelings of inferiority. A person with low self-esteem will respond more negatively to stress than a person with high self-esteem. To cope with oversized goals and low self-esteem, the individual may turn to drug abuse.

Behavioral or conditioning theories

Behavioral and emotional responses can be learned and unlearned. Reactions to behavior can reinforce the behavior or have a punishing effect. Important in this view is to question what is reinforcing the addictive behavior. In therapies the goal is unlearning the addictive behavior and replacing it by other behavior. Both classical and operant conditioning factors play a central role according to Wikler (1973) in the process of addiction and relapse. Two definitions, pharmacological reinforcement and direct reinforcement are central in Wikler's theory. He was convinced that narcotic-induced euphoria and fear of aversive withdrawal states were not sufficient to account for addiction. There is an interaction with other sources of reinforcement and he called this direct (not drug-coupled) reinforcement. The fact that there is a rapid tolerance for the euphoric effect of the drug -the addict only wants to feel normal- and that fear of abstinence is not totally realistic, especially as many heroin addicts experienced rather painless withdrawal by methadone substitution, support his view that there are other factors conditioning the use of drugs. Another fact supporting his theory is the onset of withdrawal symptoms in addicts when after detoxification they returned to environments where they had previously taken drugs.

In the view of behavior therapists, attitudes and social norms are regarded as the main determinants of the intention to produce a certain type of behavior. In the therapy it is explained to the client that he is responsible for his own behavior (Stoop et al.; 1987). The goal of the therapy is to unlearn the choice of addictive behavior in a certain situation associated with drug or alcohol use and to replace this by other behavior.

System oriented theories

Addiction can be seen as a result of a pathological equilibrium in a relationship or in a family system. In a relationship the addict creates a possibility for the partner to be the strongest of the two. In a family system the addict distracts the attention from other problems and keeps the family together by becoming a common focus of their attention.

With adolescent drug addicts there seems to be a 'leaving home' problem. Intense fear of separation occurs from both sides as the child attempts to maintain close ties with the family as well as prepares to leave the parental home. Drug addiction is a solution for getting a pseudo-independence, a pseudo-individuation (Stanton and Todd, 1982). Van der Kolk (1992) states that early traumatization by parents leads to a „negative bonding". Parents stay the most important persons when the children grow up. The children do not develop normal peer relationships and they have difficulties in leaving the parental home. Addiction has been found to be associated with early childhood separation and overprotection by the parents (Tennant and Bernardi, 1988) and emotional rejection during childhood (Kaplan, Martin and Robbins, 1984). Both over-involved, as well as detached, neglectful parents have been found in families with addicted children (Kaufman and Kaufman, 1979; Steffenhagen, 1980). Parents rarely reward acceptable behavior by the addict, while giving strong negative attention to negative behavior (Reilly, 1976). A lack of boundaries is found between generations (Alexander and Dibb, 1975; Klagsburn and David, 1977). Parents often give inconsistent or double messages to their children and do not set clear limits to negative behavior.

In the families of most male heroin addicts there is an overprotective permissive mother in combination with a passive, emotionally absent father (Fort, 1954; Rose, Battjes and Leukefeld, 1984). Stanton and Todd (1982) describe a pathological homeostasis in the families of addicts. Often drug addicts who have been successfully treated for their addiction relapse and return to their parents home when a crisis in the family has occurred (Stanton and Todd, 1982). Cancrini et al.,(1985) constructed typologies of families of drug addicts (see Part I Chapter 6). Family and partner relation therapy for addicts is based on the systems theory of addiction.

Social theories

Drug abuse can be seen as a result of a dysfunctioning society. People use drugs to escape from the pressure of society or as a protest against the norms in the society. Drug users are stigmatized and eliminated from society. They are used as scapegoats by people who are adapted to society for things that go wrong. Some advocates of the social theory argue that it is better not to treat any individual drugs addict as this would not change the real cause of addiction: the dysfunctioning society. They argue that treatment would only stigmatize people who are neither abnormal nor patients. Zinberg (1984) stated that addiction and controlled use are interrelated. Rituals and sanctions modulated the interaction. The social environment plays an important role in leading to addiction. His view is supported by the fact that most of the American Vietnam veterans who had been addicted to heroin in Vietnam, could stop their use without problem as soon as they had returned home.

   Biernacki (1986) places the social theory within the framework of social identity and other cognitive processes that fix the addict to the social environment and provide others the potentials and obstacles for recovery. He studied the process of natural recovery from heroin addiction and found that addicts who had identified themselves with the identity of a junky were less likely to recover from their addiction without professional help than addicts who still had other identities in jobs, relationships or social roles.

The self-medication theory and other psychiatric theories

Addicts can be regarded as psychiatric patients. When they use alcohol or drugs their symptoms are diminished. Symptoms of depression or schizophrenia disappear through the drugs taken. This theory can explain why after detoxification psychiatric syndromes may appear. In the third edition of the Diagnostic and Statistic Manual of Mental Disorders (DSM III), published by the American Psychiatric Association, a division is made between substance abuse and substance dependence, the symptoms are described on the basis of strict criteria (DSM III, A.P.A., 1980). A revised description was published in the DSM III-R (A.P.A., 1987) while the fourth DSM version is currently being developed.

   Psychiatric theories are largely descriptive. McLellan developed an instrument to measure the severity of the addiction complex, the Addiction Severity Index (McLellan et al., 1980, Hendriks, 1990). In research using this instrument with dimensions in seven categories (medical, employment, alcohol use, drug use, legal, family/social and psychiatric status) an absence of a relationship between the severity of the abuse and the severity of concomitant problems, questioned the view of addiction as a progressive disease as described by Jellinek (1960).

The tendency of relapse after prolonged periods of abstinence led to the concept that addiction is a lifelong chronic disease as characterized by Alcoholic Anonymous: once an addict, always an addict. As a consequence of this concept of addiction, as a chronically relapsing disease, Marlatt (1985) developed a relapse prevention program. However, maturing out of heroin addiction, when drugs cease to have their adaptive function required in earlier years, has also been described (Winick, 1962; Biernacki, 1986; Swierstra, 1986). In a follow-up study Vaillant (1973) found that after 20 years 35 to 42% (depending on the definition) had achieved stable abstinence. On the other hand, in a study by Harrington and Cox (1979) following up a cohort of 51 addicts, only one was found drug-free or abstinent, so for this sample, the maturation theory does not hold.

The adaptation-disease debate

From the multiplicity of theories it is difficult to conclude that addiction is a disease and therefore a medical problem. Critical attention to the disease model has been recently intensified in the debate of the adaptation model versus the disease model. From the 1930s, research and the treatment of alcoholism had been dominated by the disease concept, which thanks to the efforts of Alcoholic Anonymous, gradually replaced the moralistic view of alcohol addiction. It was Jellinek who described alcoholism as a progressive disease with different phases (Jellinek, 1960).

In the disease model persons can become susceptible to drug addiction. The susceptibility is generally attributed to a genetic or metabolic defect or to psychological damage occurring in childhood. When a susceptible person is exposed to the substance and at the same time, to environmental stress, drug addiction is likely to result, leading to problematic behavior. In the 1960s the disease concept began to be criticized and addiction was regarded more and more as being initially a behavior problem (Schippers, 1984). In the opinion of Schaap (1987) much of this criticism has to do with a narrow somatic definition of the medical model and alcoholism. Van Dijk (1976) responded to the criticism by describing the criteria of the illness. He describes the diagnosis, the etiology, the prognoses, the therapy and prevention of the disease.

The view on alcoholism and other addictions as a disease has been beneficial for the addict in facilitating treatment, as well as neutralizing the view that the addict is fundamentally immoral and of weak will. On the other hand it provides the addict with an excuse for his self-destructive behavior and may reinforce the attitude that he is not responsible for his situation. The disease model has provided justification for violent action by police and the military against drug traffickers in a war that seems to have no results at all. Describing addiction in medical terms or in classification systems as the DSM III is not based on etiological theories and does not give therapists guidelines for treatment. The descriptions may be only part of the problem. Instead of labelling an addict as a patient, Trimbos (1980) speaks of risk-taking deviant behavior. Schippers (1984) speaks of life-style problems.

Addiction problems are not solved by prescribing drugs to addicts as practiced in 'low threshold' methadone programs, since these make no demand to stop taking drugs (Driessen, 1987). Even prescribing heroin as was done in Great Britain did not stop addicts from continuing taking additional drugs (Denham, 1978). Clients of the Amsterdam methadone program started to use the money they had collected by stealing or prostitution to buy other kinds of drugs such as cocaine (Kooyman, 1984a, 1986). Jansen and Swierstra (1983) found in an investigation on heroin addicts in the Netherlands that more than half of the addicts in their sample had been involved in criminal behavior before they started to use drugs. The use of heroin seems to fit into a particular life-style (Van de Wijngaart, 1987).

Alexander (1987,1990) described addiction as a way of coping with difficulties. In his adaptive model the persons with a high risk of addiction are those who have failed to achieve the generally recognized levels of self-reliance, competence, social acceptance and self-confidence, that are the basic expectations of adulthood. They have failed to grow up. Addiction provides some kind of identity and purpose in life, while they have no alternative. Rather than seeing drug addiction as a cause of many problems in the adaptive model it is seen as the result of many problems in the same way as one can consider gambling, criminality, excessive use of food, television and sex as being the result of an inability to cope with the problems of life. In the disease model, susceptibility suggests a passive vulnerability to addiction. In the adaptive model the addiction is seen as a choice in searching for an organizing principle in life when normal motivation is derailed. Supporting the adaptive model is the fact that exposure to drugs does not necessarily result in addiction. The adaptive theory does not however, explain the progressive self-destructive nature of severe addiction (Alexander and Hadaway, 1982).

Another view about choosing addiction rather than coping with stress is the sensation-seeking theory (Zuckerman, 1970, 1986; Platt, 1975). This theory can explain well why some people use drugs, but not sufficiently why they cannot stop this behavior. From an adaptive point of view punitive measures directed at both traffickers and users are counter-productive. Successfully prohibiting the drugs would only force the addicts to choose another substitute adaptation. Punitive treatment of addicts can only exacerbate their sense of failure, and is therefore likely to increase their need for addiction.

Interactive models

Not one of the above described theories can fully explain the existence of addiction. They can however explain part of it; all views are valid in their own way; they focus on different factors, which can to a larger or lesser degree play a part in the origin of addiction in a certain person. These factors are also interactive. An interactive model of drug-dependence was developed by the World Health Organization working groups on „Nomenclature and Classification of Drug and Alcohol Related Problems" (Edwards, Arif and Hodgson, 1981). In this model, dependence is considered as a psycho-physiological-social syndrome, determined by a complex system of reinforcements. It not only explains how a person can become an addict but also points to the factors that keep the addict in his addicted situation.

Van Dijk (1979, 1980) gave an excellent description of the forces keeping a person addicted. In addiction considered by him as an autonomous, self-continuing harmful process, there are factors which operate as vicious circles: the pharmacological, the psychological, the social and the cerebro-disintegrative circle. He initially used this concept for alcohol addiction, but later transformed the principle to other addictions.

The pharmacological vicious circle means that as a result of stopping to use the substance, abstinence symptoms may occur. These symptoms disappear when the addict uses again. There is also a psychological vicious circle. Feelings of shame and guilt on being an addict may lead to the continued use in order to diminish these feelings. Then there is a social vicious circle. The addictive behavior leads to conflicts. Society stigmatizes and isolates the addict; subcultures develop and this makes it difficult to escape from the addict-identity which the user finally assumes. The cerebro-disintegrative vicious circle is especially apparent in alcohol addiction; excessive use can impair the integrative and regulating functions and because of this, the individual has less resistance to his craving to use again. The concept of the vicious circles helps us to understand what keeps the addict addicted. For the treatment of addiction this is of even greater importance than its etiology.

Using drugs or alcohol dependently is described by Uchtenhagen and Zimmer-Höfler (1981) as being caused by a disturbed balance between the following factors:

1. outside pressure
2. support from the environment
3. the autonomy of the individual

When one of these factors dominates the others or if one of these factors is not sufficiently present, the individual experiences stress.

The use of a drug can temporarily diminish these feelings of stress. The addict has found a surrogate solution for his problems. He feels good despite the threatening, boring or painful experience of personal, interpersonal or social origin. He becomes trapped in the vicious circles of the addiction: his only remaining problem is how to get hold of the drug in order to feel good or just feel normal. Provided the drug is easily obtainable and its use does not cause additional problems such as a poor physical condition, lack of money, or the threat of being arrested, a motivation to stop using the drug will generally not occur.

Negative and positive motivation

In most cases external factors will lead to a decision to ask for help. Very often however, the addict will end up in playing a game with the physician or the social worker in order to continue his addiction. If external factors are a real burden, the addict may ask to be admitted to a treatment center. This can be called a negative motivation. We can consider the admission as a first necessary step to treatment, a step that is not easy to take. Stopping the use is only a first step into treatment (Kooyman, 1985b). A positive motivation 'the wish never to return to drug use' usually develops not earlier than after a few months of treatment (Kooyman, 1975c).

Concepts used in the therapeutic communities

The therapeutic communities have developed from experience and have made their clients active members in their own therapy and that of fellow-residents. Both the therapeutic communities for addicts, originating from the self-help movement in the U.S.A. and the therapeutic communities for psychiatric patients, that developed in Europe after World War II, have in common that they are a reaction to the punitive care, custody and control mentality of treatment institutions (Bratter, 1985; Goffman, 1961 and Szasz, 1970). The therapeutic communities for psychiatric patients and the therapeutic communities for addicts have originally developed independently of each other. In the European therapeutic communities based on egalitarian and democratic principles described by Jones, the patients were given respect and authority by the professionals, who stepped down from their thrones and shared their opinions with the patients – now called residents -in open discussions. What happened during the day was reviewed in group meetings at the end of the day. The main therapeutic element became learning by feedback from actions called social learning in social interaction (Jones, 1953, 1983, 1984).

The therapeutic communities of the Jones' model proved to be largely ineffective in treating addicts (Kooyman, 1975a,d; Kooyman and Bratter,

1980; Schaap, 1977, 1987). The democratic structure requires that members function on a more or less adult level; most addicts who are admitted to therapeutic communities are not capable of doing so. Moreover inability to function on an adult level may well be a cause for their addiction.

In the U.S.A., treatment of heroin addiction had been not very successful until the emergence of Synanon in 1958 and was dominated by the psychoanalytic approach. According to Bratter (1985), Mowrer was one of the first authors rejecting the psychodynamic model. He believed that behavior should be considered as a manifestation of irresponsibility rather than a disease and concluded that psychodynamic therapy was 'non-therapeutic'. Psychoanalysis subtly relieves the patient of their responsibility for their acts and behavior. In Synanon the addicts were given the opposite message; they were themselves responsible for their problems: „Nobody else than you yourself put that needle in your arm" (Bratter et al., 1985a). In the American self-help therapeutic communities the addicts were given respect and authority by treating them not as criminals or patients but as persons that had been stupid to choose for addictive behavior, and by telling them that they could act responsibly and take an active role in their rehabilitation, helping themselves as well as others. They became the only treatment organization where former clients were expected to be able to become staff members and even the directors of the organization.

The concepts of Synanon

Some months before Charles Dederich had founded Synanon, where both alcoholics and drug addicts were welcome to live a drug-free life together, the AA-movement had formally excluded drug addicts from their membership. In the AA-groups held in his home, Dederich had developed a technique which was called „the Game", a leaderless group-meeting where it was safe to say anything without editing. These group meetings became an essential element of Synanon. During these meetings the group centered on one person at a time for about twenty minutes and then moved on. The norm was to attack aggressively a person who was on the 'hot seat' and not to defend this person, to break all contracts; i.e. tacit or deliberate collusions with someone to conceal material during the game. The composition of the group was continually shifted to maximize variety input (Deitch and Zweben, 1979).

Dederich, a great charismatic and qualified leader had broad philosophical interests (Freud, Thoreau, Bhudda, St. Thomas, Plato, Emerson). Ralph Waldo Emerson was often cited („There is a right way to do everything, even if it is just boiling an egg") and his essay on Self Reliance was for Synanon members compulsory literature as well as his essay on Power (Patton, 1973). From the start of Synanon, not only former alcoholics but also former drug addicts were members of the community. In Synanon the following ideology was developed: Addicts are irresponsible individuals. They used drugs or alcohol to escape from the frustrations and tensions of daily life. To be successfully treated, a re-education is necessary to learn new values and attitudes. This ideology was based on:

- the belief that each individual had a potential to grow
- the necessity for each person to develop his creative potential
- the notion that a person has to take the decision to grow and change himself; others can only assist by being a role-model, by showing concern and involvement
- the goal to live together is happiness (Broekaert, 1976).

In later years Synanon called itself a religion, mainly to qualify for tax exemption. Before this was established the Synanon prayer was written:

> „Please let me first and always examine myself, let me be honest and truthful, let me seek and assume responsibility, let me understand rather than be understood, let me trust and have faith in myself and my fellow men, let me love rather than be loved and let me give rather than receive." (Garfield, 1978).

The game remained the main tool to help the individual change. It was also a situation where interpersonal conflicts could be worked out in the openness (Simon, 1974). Self-disclosure was stimulated: make your life public. The basic philosophy of the game found its source in the AA's 4th step: to take a fearless and searching look at oneself (Deitch and Zweben, 1979). The game developed within Synanon from aggressive attacks on negative behavior to discussions with a lot of humor and exaggerations to break through a participant's defense. The games were 'played' in a more relaxed atmosphere, although there were still periods of attacks and confrontation (Broekaert, 1976).

Kahlil Gibran was one of the philosophers that influenced Synanon. A passage from his book 'The Prophet' was frequently quoted: „And a single leaf turns not yellow but with the silent knowledge of the whole tree." The sociologist Yablonsky, who is also a psychodrama therapist, spent some time in Synanon and wrote a book about it: 'The Tunnel Back' (Yablonsky, 1965). Maslow, the humanistic psychologist who developed the concepts of levels of personal needs and self-actualization visited Synanon several times in the early sixties and sent some manuscripts to the founder, which were used as material for discussions in seminars at Synanon. The visit of Casriel, who also wrote a book about Synanon, 'So fair a house', led him to structure Daytop Village in New York, a therapeutic community in New York, modelled after Synanon (Casriel, 1963). Two ex-Synanon members, David Deitch and Ron Brancato became directors in Daytop Village after the program's chaotic first year; they succeeded in setting up a therapeutic community led by para-professionals with Synanon's self-help philosophy.

The concepts of the self-help therapeutic communities

The self-help therapeutic communities in America developed concepts that had their origin in Synanon. These concepts are known to all older residents and staff. They are discussed in seminars for all residents and taught to newcomers. They represent the cornerstones of the treatment philosophy. These basic concepts provide a set of guiding principles for the therapeutic community program. The main concepts will be described here.

The choice

To come to a therapeutic community the addict has to make a choice. This choice may be between serving time in prison or entering a therapeutic community. The doors are open and the resident can leave at any moment. Before entering the therapeutic community usually an interview led by a staff member is set up and attended by other staff and some of the community residents. Before the interview the resident may have been seated on a prospect chair in front of a wall with the written philosophy of the therapeutic community. The newcomer is asked why he wants to stop using dope. His behavior of using dope is labelled as stupid. Usually the person is asked to make some investment to show his sincerity to enter the therapeutic community. He may be told to ask for help as loud as he can. The interview is a psychological preparation for the step the addict is going to take of entering into a culture almost totally oriented towards changing his values, attitudes, behavior patterns and his self-image (Sugarman, 1974).

An important element of this interview is that most of the persons interviewing him are former addicts. They are not punishing him for his addiction or showing pity towards him; they force the addict to take a look at how he acts and at the consequences of his actions. When the prospect can convince the interviewers that he needs help, he is accepted and will be hugged by everyone.

An addict is an emotionally immature person and has to grow up

Addicts are seen as immature, irresponsible individuals, unable to postpone the gratification of their needs. They can learn in the therapeutic community to use their potentials to grow and develop emotionally, physically, spiritually, intellectually, and sexually, and to use their creativity.

An addict can help himself by helping others

In being a role model for newcomers, the resident reinforces his positive behavior. This increases his self-esteem. The addict needs the support of his peers to change: „You alone can do it but you cannot do it alone."

Honesty pays

In having no secrets from the other members and sharing all his thoughts, the resident can mirror himself in the eyes of others. Honesty is rewarded by the members of the community. When you know about negative behavior of yourself or of other persons, and you keep silent, it is regarded as a negative contract with yourself or with other persons. A person who is not honest is not free, he is accumulating guilt feelings that in turn may lead to drug use. This is summarized by the principle of truth, regularly read to members of therapeutic communities at morning meetings:

> „Here we work from what we call the Truth Principle.
> One could say, that part of our therapy is truth and honesty.
> We have nothing to hide or to twist our way around.
> We are not proud of our behavior in the past,
> but at the same time we can not remain paralysed by our guilt feelings.
> Most people can get by with small lies and being dishonest to each other.
> We found, that we can not allow ourselves to do this.
> Every departure from the truth could be fatal to us.
> If we can not be straight to each other, we add more guilt to our already heavy burden.
> This gets worse and worse and could mean, that we return to dope,
> die of an overdose or end up in prison."

Guilt sharing

Guilt feelings not shared may be a reason to act out negatively in a self-destructive way. Contrary to some psychiatric theories, in the therapeutic community guilt is not seen as a symptom of depression but as a natural alarm that should not be denied or rationalized away. Hiding something may cause guilt feelings that may result in the person leaving the community. When all thoughts and feelings are communicated, the person becomes a free individual.

In the early Christian communities the members were supposed to openly confess their mistakes in the presence of their fellow community members. This was called 'exhomologesis', openness about what happens within you in front of everybody (Mowrer, 1977; Glaser, 1977; Broekaert, 1990). In the

Christian church the confession in front of all the members has become confession to a priest, in the Protestant Church in prayer to God. This confession of what you did wrong produces a great feeling of relief. In our modern society as opposed to primitive societies where open confession to the tribe-members is seen as curative, this opportunity to get relief has almost disappeared. In the therapeutic community confessing what you did wrong 'copping to guilt' is a major healing element in the treatment process. It can take place during morning meetings when questions have been raised on who was responsible for something which went wrong, or during an encounter-group or at special sessions of the whole community in which the staff invites the residents to share guilt feelings, named 'copping sessions'.

The destructive process of not sharing guilt feelings is illustrated by the concept of the guilt circle (Fig. I). The guilt circle can be explained as follows: A person does something negative. He feels guilty, denies the feeling, finds excuses and rationalizes the feeling away, but this is not working. He is going to show positive behavior as a compensation, but he remains tense and gets afraid people may find out what he did wrong. He starts to distrust other persons and assumes that they are angry with him. He feels bad about himself, and gets depressed. Vague feelings of guilt appear no longer related to a specific negative behavior. This becomes unbearable so something negative is done to be able to attribute these vague guilt feelings to some-

Figure I. The guilt circle. After negative acting out the defense mechanisms such as rationalization, compensation and denial are used unsuccessfully, leading to fear, distrust and depressivity with vague unconnected guilt feelings resulting in new acting out behavior.

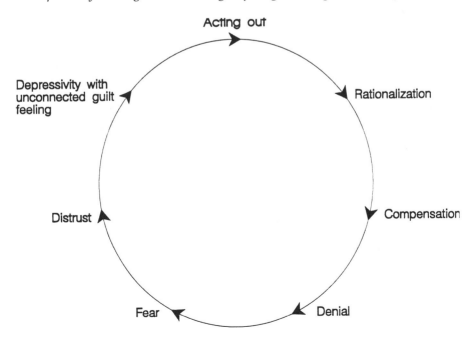

thing obvious. Then the circle may start again. The only way to stop this, is to share the guilt feelings by confessing the negative behavior and facing the possible consequences. Addicts have been denying guilt feelings for a long period. When they allow themseves to feel guilty of something negative, they become overwhelmed by guilt feelings related to negative behavior in the past. In the therapeutic community these feelings can be expressed and dealt with in the groups. The confessiom of guilt in the presence of other residents may be an important binding factor in the community (Cassuto,1981).

100% performance

In contrast with the junky lifestyle, the resident in the therapeutic community is expected to be clean; not only in not using drugs, but also in keeping the house clean and taking good care of his personal belongings. He is expected to be exactly on time and high demands are placed on the quality of his work. These high demands teach the resident to cope with feelings of frustration and depart from the laissez-faire life-style he was used to.

Assuming responsibility

The resident is made aware that he is responsible for his actions. He is ultimately responsible for his addiction. Nobody else put the needle in his arm but he himself. In the therapeutic community he is made aware of the impact of his behavior, he is shown the consequences of his acts and is told to be responsible to himself and society (Bratter and Hammerschlag, 1975).

Act as if

Residents are simply told to act as if they are responsible and by acting they become just this. They are told to act as if they are not afraid to do something and by doing so, they lose their fear. They should act as if they are full of energy, as if they understand the values of the therapeutic community, act as if they are the person they want to be: by doing it they will gradually come to think and feel the way they are acting and thus become that kind of a person they want to be. The act-as-if concept is also important by giving information to newcomers. The older member of the community doesn't have to be as positive as he comes across to the new member (Sugarman, 1974).

This act-as-if rule is explicitly not valid during the encounter groups, except maybe in the situation where someone acts as if he is not afraid to confront somebody with their behavior. In the analogy of changing how you feel by changing your thoughts, which is one of the principles of Rational Emotive Therapy, a person can change how he feels by changing his behavior (Ellis and Grieger, 1977).

Showing concern

Residents are expected to show concern for other residents and for what they are doing in the community. It should not be mistaken for being kind and helpful, taking away the responsibility for someone's acts (sympathy kills the dope fiend). Out of concern you can angrily tell a person what you think of him and confront the person with his destructive behavior. This is called 'responsible concern' (Ottenberg, 1978; Waldorf, 1971). The new member has to learn to understand this expression of 'tough love' (Cassuto, 1981). He learns that angry confrontation is not a rejection of them as a person, but a way of showing concern with their negative behavior.

Keeping the environment drug-free and violence-free

The addict enters a community with a few cardinal rules: no use of drugs, alcohol or any other mind-altering substances and no violence or threats of violence. The resident cannot act out in taking drugs or alcohol or using violence. The drug-free treatment is directed at the problems that are the reason for the abuse and they can only be dealt with in an effective way if the use of the addicting substance has been stopped. This can be done in a separate detoxification center, applying methadone or other chemicals to alleviate withdrawal symptoms or without these drugs (cold turkey). This can also take place within the therapeutic community itself (Kooyman, 1980).

Violating a cardinal rule in the therapeutic community means expulsion. This does not mean for always. Within a week, the person may be re-admitted after being thoroughly interviewed on his commitment. In an environment where acting-out in a negative way is not possible, the resident learns in a positive way to respond to stress. This can be described as follows (see Fig. II.1 to Fig.II.4:

Figure II.1: X feels good, uses D (heroin), does not feel the painful problem, does not solve the problem, also does not move away from the problem.

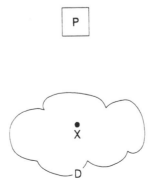

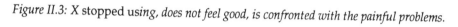

Figure II.2: X still feels reasonably good. He gets substitute drugs (methadone), does not feel the painful problem, still does not solve the problem, nor is moving away from it.

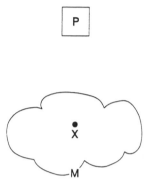

Figure II.3: X stopped using, does not feel good, is confronted with the painful problems.

Figure II.4: X is in a drug-free therapeutic community and learns with the help of a fellow resident Y to deal with the painful problem.

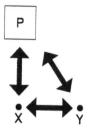

P = problem, X = addict, D = drug, M = substitute drug, Y = fellow resident

The addiction is a symptom for underlying problems

Addiction is seen as the result of a problem. It may have started as an attempt to solve a problem, but in time it became a problem in itself. When the symptom of the addictive behavior is removed, the resident learns how to find out what the function of his addiction was, and what the pain or the fear of pain was that had to be anaesthetized.

Mandatory active participation in the therapeutic program

The program cannot be consumed in a passive way. The addict has to take an active position in the community; it is „grow or go" (Jongsma, 1981). The therapeutic program is in principle the same for everybody. The resident cannot refuse to take part in any group activities. After the resident has learned to understand the program, its values become internalized and the resident starts to feel responsible for the rules, norms and values of the therapeutic community. In doing so, the resident himself becomes a role-model for newer residents.

The above described concepts are the most important concepts used in the therapeutic communities for addicts. They are often in an implicit way part of the philosophy of the program.

The the treatment philosophy of the therapeutic communities

A great advantage of the concepts of the self-help therapeutic communities is that they are easy for the residents to understand. As the philosophy of their program, most therapeutic communities have adopted the philosophy written by Richard Beauvais when he stayed in Daytop Village:

> „We are here because there is no refuge, finally, from ourselves.
> Until a person confronts himself in the eyes and hearts of others,
> he is running.
> Until he trusts them to share his secrets,
> he has no safety from them.
> Afraid to be known, he can neither know himself nor any other.
> He will be alone.
> Where else but in our common ground can we find such a mirror?
> Here, together, a person can at last appear clearly to himself.
> Not as the giant of his dreams,
> nor the dwarf of his fears,
> but as a man, part of a whole, with his share in its purpose.
> In this ground we can each take root - and grow.
> Not alone anymore, as in death,
> but alive to ourselves and to others."

In many therapeutic communities in many countries this philosophy is read at morning meetings and discussed in seminars. Having a philosophy that is discussed by residents and staff members, is in itself an element of great therapeutic value (Ottenberg, 1991).

The relationship between the theories on addiction and the treatment in therapeutic communities

How does the treatment in these therapeutic communities relate to the existing theories on addiction? First of all a failure to differentiate clearly between use of drugs and addiction to them has created confusion in etiological discussions. Use becomes addiction if a person has lost control of the drug. Instead the drug has taken control of his life. The addict is no longer free to use or not to use. Not using becomes an unbearable situation and using becomes an obsession.

Addiction can be defined as follows: Addiction is a self-continuing harmful process resulting from adaptive behavior that went out of control, becoming a problem in itself.

Why it seems to be self-continuing can be illustrated by the effect of the vicious circles of different factors. Using Van Dijk's concept of the vicious circles of addiction (van Dijk, 1980), the following circles can be drawn (see Fig.III):

Figure III: The vicious circles of addiction.

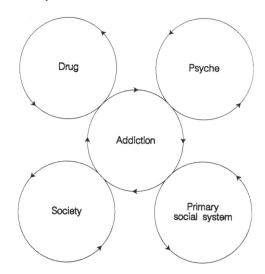

The addict gets trapped in a system of vicious circles each of them may play a more or less apparent role. As long as the drug is readily available and no additional secondary problems occur, such as bad health, lack of money, threat of being arrested, family pressure to stop, the addict does not usually show any motivation to stop, because when he stops the following effects can be seen:

1. In the pharmacological circle a discontinuation of the use causes an effect that is opposite to the desired effects of the drug: thus by starting to use again annoying, undesired effects disappear.
2. In the psychological circles the addict feels guilty because of his drug abuse. The unpleasant guilt-feelings disappear, when he uses the drugs again. The negative feelings which were repressed by the use and which were the very reasons for using initially return even more intensely than before. They disappear when the addict resumes his use.
3. In the primary systems circle: The addiction serves to sustain a pathological balance. In a family the addict serves as a scape-goat. The addiction problem distracts the attention from existing problems in the family. When the addict stops using, the other existing problems become apparent, resulting in the re-assumption of the scape-goat role by the addict. In a relationship a person may be unable to cope with a strong partner whose own self-esteem is boosted by staying with a weak addicted person. In a group of addicts somebody who wants to stop is discouraged as this may awaken guilt feelings in the rest of the group with relation to their own use.
4. In the social circle: society stigmatizes the addict; this stigmatization leads to rejection and isolation of the addict. This makes the integration in society of the addict difficult, after he has stopped his use. The addict feels only accepted in the addicts' world. Outside of this world he is seen as: once an addict, always an addict.
5. In the cerebro-ego weakening vicious circle the excessive use may have a damaging effect on cerebral functions, that regulate and integrate the behavior. The outcome is a weakening of the ego-strength. The individual psychological power to regulate and control the use is reduced. This is especially the case when drugs have a harmful effect on the brain as is the case in heavy alcohol consumption. Due to the loss of ego-strength, the motives to use become increasingly difficult to resist.

To successfully stop the addiction, all of these vicious circles have to be eliminated and not just some of them . This can be done in the 24-hour drug-free environment of the therapeutic community (Kooyman, 1986a). By offering a drug-free environment without the distribution of psychopharmaceutical drugs, the pharmacological vicious circle is excluded. The influence of the primary system to which the addict belonged is excluded by removing him from this system. This in itself has consequences for that system. When the addict is expected to return to this system or at least to have contact with it – as is the case with partners, parents or other family members – it is necessary to involve this system in the therapy, in the hope that they will be able to cope with a recovered addict after his discharge from the therapeutic community. The social vicious circle is excluded by removing the addict temporarily from society. In the new society of the therapeutic community, the resident is no longer isolated for having been addicted, instead he becomes actively involved in this micro-society. It is a challenge for therapeutic community programs to educate society to accept the possibility of treating addicts successfully. In this way they can change the assumption: „once an addict, always an addict." By excluding four of the five vicious circles, the focus during the time the addict spends in the therapeutic community is on the psychological factors that play a role in the relapse into drug abuse. Due to strict social control within the therapeutic community, drug or alcohol use is impossible in the house, so if a person resumes the old habit, he has to leave.

The focus on the individual: the psychological vicious circle

In the treatment the focus is on the individual and on his interaction with other members of the community. There we have the situation that the addictive behavior, the symptom, has been stopped. Let us now look into the theories on etiology of addiction described earlier in respect to what happens to the individual during his stay in the therapeutic community. The same factors that may have led to addiction as a response to coping with a given situation may cause a relapse into addiction after the addictive behavior has been stopped for a certain period.

So why do people use drugs in such a way that they become dependent on the drug? Are the theories describing potential etiological factors that play a role in the development of a state that we call addiction, applicable to the therapy of the therapeutic community?

The biological theories on addiction and the resident in the therapeutic community

A person that gets addicted may be biologically more vulnerable than others. This vulnerability may be inborn or acquired through traumatic experiences in early childhood. People who become addicts may have a decreased tolerance to pain. The therapeutic community program offers no substitute drugs. Substitution of one addiction for another addiction keeps a person dependent not only on chemicals but also on providers of these chemicals, which is contradictory to the goal of the program.

In the program, especially in groups, the residents learn to experience painful emotions. They learn to express them rather than hide them. It is unknown if through these emotional groups, for instance in the new identity process developed by Casriel, biochemical reconstruction of damaged patterns occur. In emotional groups, closeness of other persons both physical and psychological, evokes pain from the past which has stored away. This is according to Casriel, pain of early deprivation, of being rejected as a small child. A biochemical dysfunction as a result of early traumatization may partly explain the avoidance of physical contact and fear of intimacy seen by addicts. In this way, the addict avoids rejection and the accompanying pain (Kooyman, 1991).

Now we also see the opposite form of behavior, a tendency by addicts to seek painful experiences, as if, in a neurotic way, trying to master the painful situation. In this way they follow the pattern of persons severely traumatized, who show a tendency to repeat in different ways the traumatic situation. In real life this rarely seems to be a self-healing therapeutic activity. However, in therapies which use techniques to express emotions as Casriel's New Identity Process, or Janov's primal scream therapy, screaming is used as a tool to express emotions while emotionally reliving traumatic events of the past (Casriel, 1972; Janov, 1970). Also, other therapies go back to the traumatic situation such as for instance psychodrama (Moreno, 1959), hypnotherapy (Van der Hart, 1988) and sessions using LSD or Pentothal to break through the individual's defenses against painful memories (Bastiaans, 1986). These therapeutic interventions seem to work with traumatized people. An unanswered question is if they work at a biochemical level in the body. If they do, the therapies used in therapeutic communities such as encounter groups may do so too. Since sufficient research is still lacking, this is, however, highly speculative.

Psychodynamic theories on addiction and the resident of the therapeutic community

The therapy in the therapeutic community can be looked upon from the point of view of psycho-analytical theories. The therapy differs in many ways from the traditional analytic therapy. For instance, the person first has to change his behavior, then experience the emotions that accompany this change and then he has to learn to understand what is happening. In psychoanalysis it is usually the other way round. The therapist in the therapeutic community is emotionally involved and expresses his emotions, uninhibited by the concepts of transference or counter-transference. Inborn negative instincts are denied, negative behavior is seen as reactive to situations. A person is seen as not having been born bad and is expected to be able to grow up to self-actualization by developing his potentials.

As in analytic therapy, in the therapeutic process defenses appear as a reaction to the pressure of the therapeutic system. When a resident is under pressure of the treatment program, acting-out behavior usually appears. This behavior can be understood as an externalization of internal conflicts, and seen as originally adequate but after the person has grown up as a defective way of expression, as a defense and compensation. An example is a person hurting the persons he is interested in; evidently anticipating that these persons might not be interested in him and therefore preventing himself from getting into a situation where rejection could take place (Uchtenhagen, 1985b).

Most of the addicts admitted to a therapeutic community are of the character disorder type. Before they used their first drug, the behavior displayed, served as a protection against often unconscious, painful feelings. Taking drugs helps to attain the same goal: the drug use can be described as acting-out behavior. It is part of an already existing defense mechanism. The most important defense mechanism of the character-disordered person is projection. Negative feelings of the addict are projected on to other persons. Acting-out, described as an externalization of internal conflicts, can be seen as a defense. This defense can originate from a defective way of expressing or facing the inner conflicts due to the inability to endure the accompanying emotional stress (Kooyman, 1985a).

Acting out behavior is characterized by the absence of direct emotional response. Anger is not shown directly to the person provoking the emotion. Love is not shown directly either. Instead the emotion is acted out towards other persons or objects, while different emotions may be directed at the person provoking the emotion. For instance, to a person that provokes love, fear or anger is shown. To a person that provokes anger, love may be shown or no emotional response at all, while the anger is directed to other persons or objects. Acting-out can be seen as a conflict, as avoiding behavior, as an expression of an inner conflict or as a compensation, protecting narcistic vulnerability by narcistic upgrading (in community terms building up a big image). Acting-out behavior can be regarded as a defense against being hurt. A major fear is the pain of being rejected (Kooyman, 1986a). The person often, however, provokes the rejection by his own behavior. When he provokes the rejection in a clear way himself, the rejection becomes bearable (Kooyman, 1985a).

The therapeutic environment of the therapeutic community provokes acting-out behavior but also sets limits to acting-out behavior. A therapeutic setting can provoke acting-out behavior to an uncontrollable extent as was the case in the early months of the existence of the Emiliehoeve therapeutic community (see part II). The most important issue is to find a balance in an environment that offers enough pressure to provoke acting-out and at the same time provides enough structure that sets limits to prevent uncontrollable acting out behavior. In order to grow, the resident has to drop his defense mechanisms. The residents in the therapeutic community call these defenses 'images, hiding the real self'. When the defenses are removed the resident feels vulnerable and may think of running away from the pressure. Usually after some weeks or several months, a regression takes place: 25 to 35 year old residents regress to function at the age-level in which all went wrong. They behave like teenagers and act out their authority conflicts with staff and older residents. In the therapeutic community they grow up again in a healthy way to be able to leave their new home, hopefully for good, as grown-up adults.

Residents have almost without exception a low self-esteem. Alexander Bassin (1980), one of the founders of Daytop described addicts as follows: „Addicts are persons with a low self-esteem they have a failure identity and are unable to sustain prolonged emotional relationships." Addicts have not only failed in their relationship, in their education or jobs, but when they enter the therapeutic community, also as an addict. Asking for help is the very thing an addict avoids doing and entering a therapeutic community can be considered as zero, although he may hide this idea of himself in his behavior. The very reaction of the interviewers at his admission to the therapeutic community that he was stupid to use drugs or alcohol, already increases his feelings of self-esteem. In this way the message is given that he is able to act differently and that apparently he has a choice (Sugarman, 1974). Punishment in jail or treatment as a patient in a hospital had only led to a further decrease of his feelings of self-respect. If he would have been a hopeless case, he would not have been called 'stupid'. By acting as if he is not afraid to do things and by assuming responsibility in the daily life of the therapeutic community, and after accomplishing tasks he thought he was unable to carry out, the addict's self-esteem grows.

Preston and Viney (1984) found that residents in therapeutic communities in Australia who had been in treatment for 10 months or more had higher self-esteem than residents in treatment for shorter periods. In research among residents and ex-residents of the therapeutic community Daytop Village, Biase found a low score on a self-esteem measurement list in most of the new residents. The self-esteem increased during the program with most of the residents and remained at a higher level after discharge. Those who did not show an increase of their self esteem during their stay in the program, had a much higher chance of relapse. This was found among the ex-residents who were seen in the follow-up study after discharge (Biase & Sullivan, 1985). Another interesting finding was that residents who had followed school education during their stay in the therapeutic community, were doing better on the self-esteem score than those who did not receive this education during their treatment in the therapeutic community.

The behavioristic theories on addiction and the resident in the therapeutic community

The therapeutic community is a place where addictive behavior is unlearned and other behavior is learned. The members of the community reinforce positive behavior. Being responsible for your own behavior is one of the basic concepts for behavior therapy with addicts, as well as for the treatment in the therapeutic community. „Going through the emotions and acting as if" can be explained as a behavior therapy technique. There is a continuous reinforcement of positive behavior. Negative 'junky' behavior is heavily confronted. The residents are doing many things they are afraid to do. By doing it anyway they overcome their fear.

System oriented theories on addiction and the resident in the therapeutic community

The therapeutic community is a new primary system. Therapeutic communities provide a substitute family to a population whose previous experiences with family life have often been unsatisfactory. Dederich, founder of Synanon, conceptualized the process as one in which the addict was reborn and was going through an accelerated maturation. The 24-hour support system made the stopping of maladaptive defenses and their replacement by more positive coping patterns possible (Deitch and Zweben, 1979).

A major difference between Synanon and the therapeutic communities modeled after Synanon is the decision in 1968 to make Synanon a community for life to cope with the relapse problem, while other therapeutic communities continued to view the re-entry into society as an achievable goal. Most residents after discharge go to live on their own or with a friend, at least those from North American and North West European therapeutic communities. The therapeutic community is a place where the residents stay between the phase of staying at or being in frequent contact with his family of origin, and living on his own. As the family or partner may play a negative role in whether or not the addiction is conciuosly continued, the contacts with family members, partners and friends are usually not allowed in the first months of the program. Admitting the resident has an effect on the parents, partners and other relatives. Therefore they are involved in the treatment from the moment when contact with the resident is allowed again. However, involvement of these persons begins preferably from the start of the treatment even before admission. A later chapter describes how parents, partners and other relatives can become part of the treatment program of the therapeutic community.

The substitute family offers a safe place to grow up and to learn from mistakes; mistakes are important. Residents have to learn that making a mistake is not a disaster, as they may have thought as a child. There are clear limits set in the therapeutic community to negative acting-out behavior. There are strong authority figures, a strong father figure is usually present. Positive behavior is rewarded. Negative behavior is punished. In the new family the addict learns to respect himself and other people. He also learns to give and receive love and affection.

The social theories on addiction and the resident of the therapeutic community

Addicts in a therapeutic community are in an environment very different from the places where they used to stay when they were using drugs or alcohol. The house in which they are residents is kept very clean. No reminders of the drug or alcohol culture are around. The residents themselves are stripped from reminders of the drug sub-culture. The long hair is cut, the junky clothes are burned or thrown away and replaced by tidy, nice-looking clothes, sometimes after an initial period in the community in which they have to wear overalls.

From the therapeutic community they pay visits to the 'world outside', to concerts, to museums, to sport events, to theatres. They are educated about the good and interesting things in society in seminars. In the last phase of their stay in the therapeutic community they are usually encouraged to join a club or society outside the therapeutic community to stimulate socialization with persons from outside the therapeutic community. From a participant in the drug scene they have become a responsible member of the therapeutic community. In the re-entry program their therapeutic community identity has to be changed in a personal one, no longer related to addiction.

The therapeutic environment

The treatment in a 24-hour setting meant that it takes place within a social system. Roughly there are two basically different perceptions of this idea. One is that in-patient treatment settings are total institutions with all their possible disadvantages. This perception is described, among others, by Goffman in his book, 'Asylums' (Goffman, 1961). The other perception is using the community as a therapeutic instrument. In the first view the environment is perceived to make the patient worse, while in the second view the perception is that the environment can make the patient get better. In a good climate with open communication the environment can become therapeutic: social learning can take place through confrontation, structuring, involvement and validation; factors described by Gunderson which have to be present in a therapeutic environment that is good enough (Gunderson, 1983).

The staff plays an important role in keeping the environment therapeutic (Jongerius, 1989). Even as early as in 1954, when describing the mental hospital, Stanton and Schwartz (1954) emphasized that staff conflicts have a negative effect on the climate and as a result also on the individual treatment of the patient. A good safe climate is important to a therapeutic community. To achieve a safe environment a therapeutic community needs structure with a tradition of norms. In a new program these norms have to be made clear by the staff. In older therapeutic communities things happen because they have become a tradition. In a new therapeutic community norms such as being on time for meals and meetings, knocking at the doors of staff offices, have not yet become part of a tradition and have to be structured by the staff. In existing therapeutic communities the residents learn these norms very quickly as newcomers and things happen in a certain way because they always happen in a certain way. Starting a new therapeutic community is easier when a tight structure is introduced by the staff rather than when it is left to the residents to develop norms over a long period of time.

The setting

The building of a therapeutic community is not unimportant. To be able to create a group feeling a large common room has to be available where residents and staff can meet. Privacy should be minimal for residents during their stay in the therapeutic community to prevent them from isolating themselves. The bedrooms should be for at least three persons. In a bedroom for only two persons, pairing off and making negative contracts becomes easier. In the re-entry phase, the residents should preferably be assigned single rooms.

A reception desk at the entrance of the therapeutic community where visitors can be received is important. The area around the building should allow some space for residents to walk around or to be engaged in sports. The borders of the premises must be clearly indicated so that it is obvious when one has left the premises.

The therapeutic process in the therapeutic community

The resident in the therapeutic community is an active participant in his recovery. The group process is a basic element in the therapy. Group therapy was found to be particularly effective in the treatment of persons addicted to alcohol or drugs (Kooyman and Esseveld, 1984). Within the therapeutic community the main therapeutic element is the encounter group. These meetings can be considered as an intensive form of group psychotherapy. Addicts have a low self-esteem, do not develop trust in themselves and do not trust other persons. They are afraid to be rejected and manipulate people to avoid being rejected. They are not capable of sustaining emotional relationships. These characteristics make addicts not the most suitable people for individual psychotherapy. In fact individual therapy usually fails, unless there are factors that help to control abuse outside the sessions.

The therapist who has extensive experience in working with this category, is aware of the games addicts play with the therapist, resulting in no changes taking place, especially when the therapy is non-directive. The main therapeutic difficulty is the lack of trust the addict has in other persons: „other people may hurt you and they can hurt you more when you have developed affection for them". Group therapy has the advantage that trust in the group can be gradually built up. At the start the addict may trust only some of the group members. Slowly he will grow to trust others. It helps when he can identify with group members as they have been addicted too. Identification helps to generate trust. In the groups the old games cannot be played; being experts in this manipulative behavior, the participants challenge this behavior and stop these games . The new member trusts heavy confrontation more than he trusts a friendly understanding attitude. When a person who has a very negative picture of himself is met with friendliness and affection, this usually is very scary for the new resident and often a reason to break off the contact and leave the program. Being too friendly too early to an addict who comes for treatment, is more dangerous than having a strong confronting approach (Kooyman and Bratter, 1980).

Yalom has described the curative factors of group therapy (Yalom, 1975). As group therapy is a main therapeutic element in the therapeutic communities, the curative factors of group therapies are present in therapeutic communities (Hollidge, 1980):

1. *The instilling of hope*

   Upon admission to a therapeutic community, the resident sometimes suffers from a conscious feeling of hopelessness and worthlessness. He has failed in everything, even as an addict. In a therapeutic community it is explained to the resident that he can change. Such a change is not possible without causing conflicts; if there is no conflict, there is no stimulus to change.

   It is very difficult to instill hope for a positive change in individual or in group therapy on an out-patient basis, since the client is reinforced outside the sessions in his continuing failure. In the climate of a therapeutic community, where no-one is ever seen as hopeless and the person in question is stimulated to use his existing potential, it is possible to eliminate the feelings of hopelessness and worthlessness; at least when someone in the group calls you stupid he does not make you feel a hopeless case.

2. *Feelings of togetherness*
    Confrontation with the problems of fellow residents makes the residents' problems no longer unique. Recognition of the other's situation develops confidence and trust. Contrary to the situation in individual or group therapy, the client of a therapeutic community does not feel alienated from his social life.

3. *Altruism*
    Most addicts believe that they are of no value to others. By helping others, their feelings of self-esteem are reinforced.

4. *Socialization*
    Instead of the manipulative behavior that the addict displayed to others in order to satisfy his own needs, the resident learns new communicative skills. He learns to ask for help in a direct way.

5. *Development of interpersonal skills and sharing information*
    The resident learns to express and to accept criticism. Through exchanging experiences he acqires the necessary insight into his own behavior. In this way reality-testing is improved.

6. *Group cohesion*
    New residents may see the therapeutic community as a threat and consequently isolate themselves for fear of rejection. Confrontation with this gives insight into this behavior. Feelings of belonging to the group have to be developed in order to eliminate underlying feelings of not being accepted.

7. *Re-living situations from the family of origin*
    A therapeutic community serves as a new family in which positive expectations can be experienced.

8. *Identification*
    When the first positive relationship with fellow-residents has been developed, a clearly different identification can emerge from that which the addict developed towards his parents. The old identity is left behind and a new and positive identity develops producing an increase of self-esteem, insight and trust.
    In contrast to out-patient group therapy, where the above described therapeutic factors can also occur, these elements are part of a larger therapeutic environment, which are continuously present.

Bonding therapy: its potential use in therapeutic communities

There is one specific area which is problematic for most addicts that is not dealt with sufficiently in the traditional therapeutic community, the emotional and physical closeness to others. Most addicts avoid intimacy. Physical closeness with a partner is often only associated with sex. If this fear of closeness is not dealt with in the therapeutic community, the ex-addict may be able to become socially responsible, a hard worker if not a work-aholic, but unable to sustain a stable relationship with a partner. Casriel (1972) stated that bonding, a state of being emotionally and physically close to another person is a biological need and that fulfilment of this need is necessary for having a happy life. He described two opposite reactions to the bonding of persons who had not learned as a child that they did not have to pay a price to develop bonding, that it was just unconditionally available, that of an acceptor and that of a rejector.

The acceptor will pay any price for affection, he does a lot to get it, he suffers it or undergoes even humiliation to get any affection or bonding. He is afraid to be left alone, is not able to express anger; when he gets affection without having to suffer he does not accept it, he is pain dependent. Used to paying a price for getting affection, affection just received for free is not trusted and is run away from. The rejector has learned that the price for affection is too high and has decided that he does not need anybody's affection, he cannot ask for help, he does not show his needs, he learns to rely on himself, he feels safer when alone, he cannot express fear or pain; he freezes, encapsulating his emotions. Most addicts have developed a rejector's identity, they don't show their needs to other persons, they would rather take drugs or alcohol instead.

In the bonding therapy groups, which Casriel called the 'New Identity Process', the participants learn to overcome their fear for closeness, for bonding. The experience of human closeness in the group recalls the pain of what has been denied to them as a child. Once the old painful situation of the past has been emotionally worked through, the participant can enjoy emotional and physical closeness (Casriel, 1971). In the Emiliehoeve Therapeutic Community bonding-groups have become an integral part of the program. They seem to add an important often missing element to the therapy in the therapeutic community.

The therapeutic community itself is the therapy

In the environment of the therapeutic community the resident learns to be able to respond to stress in a positive way. He loses his fear of failure and that in itself releases inner stress. The feeling of stress is experienced more by expecting a stressful situation than by being active in a stressful situation itself. An actor experiences more stress before the onset of the performance than during the performance itself. This is also the case with sportsmen before the start of a race. The resident learns to actively cope with stress and crisis situations instead of relieving the stress by taking alcohol or drugs or other acting-out behavior.

As a result the self-esteem is increased and the fear of rejection is diminished. Through the experience in the therapeutic community, the addict is able to live a life without dependence on chemical substances to relieve stress. When we consider that large therapeutic communities, housing more than one hundred residents in a system with only a handful of staff members have success with most residents that stay long enough to be affected, we must realize that the therapeutic effect is not the input of the staff. The staff often hardly know the names of the residents, the main therapeutic element is the structure and the philosophy of the community itself. The staff members are usually ex-addicts who have graduated from the program and they have a positive effect on the residents as role-models. The therapy is the therapeutic community itself.

The therapeutic factors of the therapeutic community for addicts

What is therapeutic in the therapeutic community? The therapeutic community offers a 24-hour drug-free environment. Using the different theoretical concepts described in this chapter as a frame of reference the therapeutic characteristics of the therapeutic community for addicts can be summarized as follows:

1. *The substitute family*
    It serves as a substitute family offering the resident a possibility of growing up in a safe environment.

2. *The consistent philosophy*
    A consistent philosophy that can easily be understood, explained and supported by all members.

3. *The therapeutic structure*
    The environment has a clear structure where no double messages are given. This structure offers safety and security to the resident. There are few but clear rules. The environment offers enough pressure to learn, but also enough structure to prevent chaos from developing without becoming over-organized. Mistakes must be able to be made; a therapeutic community with too many rules is dead. The resident can move in the structure of the therapeutic community to positions of increasing responsibilities at different levels of their development.

4. *The balance between democracy, therapy and autonomy*
    In the community there must be a balance between democracy, therapy and autonomy of the individual. Democracy in society is meant to be a way of delegating power to all persons, of making decisions possible by a majority and of solving conflicts. A therapeutic community delegating all power to the residents who are in the majority compared to the staff, can become anti-therapeutic and used as a collective defense against therapy and a way of avoiding conflicts. That was illustrated in the first months of existence of the Emiliehoeve Therapeutic Community (Kooyman, 1975a). The therapy in the therapeutic community is obligatory. The resident cannot choose whether or not to take part. The autonomy of the individual is limited by being part of a group.

5. *Social learning through social interaction*
    Learning takes place as social learning through social interaction (Jones, 1953).
    The resident cannot isolate himself in the therapeutic community. He has to be actively involved and continuously receives comments on his behavior and attitudes. Elder residents serve as role-models for the newer ones (Comberton, 1986). The resident learns to function in different social roles. Through feedback he gains insight into his own behavior. He learns to make choices and accepts the consequences of these choices.

6. *Learning through crisis*
    Learning is fostered by experiencing crisis situations. It can be understood by Erikson's theory of social learning. He regards maturation as undergoing a series of crises leading to disorganization, followed by reintegration at a higher level, after the crisis situation has been mastered (Erikson, 1963).

7. *The therapeutic impact of all activities in the community*
    Everything that happens in the therapeutic community is therapy. The members have to prepare the food, work in departments and have to learn to function in different roles.

8. *The responsibility of the resident for his behavior*
    The resident is not regarded as a patient. As described by Parsons (1951), the following elements can be attributed to the role of a patient:
    a. the patient is relieved of normal role obligations;
    b. the patient is not regarded as responsible for his situation;
    c. the patient is expected to see his situation as undesired and to leave the role of the patient as soon as this is considered medically justified;

d.   the patient is obliged to seek help.

The first two elements are not attributed to the resident in the therapeutic community. They are told that they should not play the victim or blame their failure to achieve desired goals in life on their addiction while blaming others or society for their situation. Instead they are told they behaved childishly and had been stupid and irresponsible. The resident is labelled as a person who needs help and who has to learn to ask for help. Also the resident is seen as being able to help his fellow-residents.

9.   *Increase of self-esteem by accomplishment*

By overcoming the fear of failing to accomplish things step by step, by acting 'as-if' he is not afraid, the low self-esteem of the resident is raised. This is also the effect of being engaged in creative activities and education classes in the therapeutic community. An increased self-esteem in itself diminishes the fear of failure and of rejection. In helping other persons the resident realizes that he is of value to others which reinforces his self-esteem.

10.   *Internalization of a positive value-system*

A positive value-system is imposed and internalized. The resident learns to be honest, to confront and criticize negative and self-destructive behavior and attitude and to see a problem as a challenge.

11.   *Confrontation*

The confrontation and pressure put on newly admitted persons should be of a limited character to avoid pushing them out of the therapeutic community. Life in the therapeutic community is in itself a lot of pressure for a new resident to take. When the resident starts to develop in the therapeutic process, trust in the program appears to grow and the resident drops his defenses. In this frightening situation, experienced as a crisis, the resident is ready to change his behavior and attitudes. The resident learns to verbalize his inner conflicts and to face the emotional stress linked with feelings of guilt and incapacity. The basic fear of annihilation disappears and the person is able to discard his defense completely. Now he takes the step towards direct emotional confrontation of other persons instead of showing acting-out behavior. As he learns that fellow-residents appreciate the new behavior, the fear of rejection begins to disappear. He learns that it is not necessary to be perfect to be loved as he might have thought as a child. In fact, the opposite seems to be the case. As Shostrum wrote: „We seem to assume that the more perfect we appear, the more flawless, the more we will be loved. Actually, the reverse is more apt to be true. The more willing we

are to admit our weaknesses as human beings, the more lovable we are."
(Shostrum, 1967). The resident learns that confrontation is not directed at
the person but at his behavior.

12. *Positive peer pressure*
In the same way that peer-pressure may have been a factor in leading a
person to use drugs in the first place, so positive peer-pressure makes a
person decide to abstain from taking drugs of alcohol and inside the
community to develop positive behavior. Residents in the therapeutic
community are confronted with negative behavior by peers. Individuals
further in the process serve as role models. The resident is made to feel
that it is his responsibility to observe and criticize or comment on the
behavior and attitudes every other resident. This is a complete departure
from the code of the streets where addicts will not disclose to the police
or any official the activities of another addict. Playing the victim is not
rewarded and manipulative behavior is quickly discovered and sharply
denounced.

13. *Learning to understand and express emotions*
In encounter groups and other therapeutic group meetings, the resident
is encouraged to express his emotions. Usually screaming is used as a
tool to express emotions in a clear way. The addict learns to overcome
his fear for expressing his anger, fear and pain, emotions that no longer
need to be defended against in acting-out behavior. Finally, he may learn
to experience positive emotions such as pleasure and love, which are
usually still more difficult to express than negative ones.

14. *Changing negative attitudes to life into positive ones.*
Most addicts have negative views of themselves in relation to other
persons, such as „I'm not lovable", „I don't need other people", „I don't
have the right to exist". That last attitude is very common among
addicts, although they often only realize during therapy groups that
they have such an attitude towards life. Their negative attitudes
developed in early childhood. It helped them to survive when they were
small children, but it became a great handicap when they grew older. It
takes time to change negative attitudes about oneself. Groups such as
The New Identity Process -in which false identities are replaced by real
identities- can help to overcome this handicap of having one's own
negative self-fulfilling prophecy on life (Casriel, 1972).

15. *Improvement of the relationship with the family of origin*
The relationships with parents and other relatives are renewed with the
help of staff members from the moment contact is allowed after an initial

period of no contact. In that first period, parents went to parent groups sometimes even before admission. In some cases family therapy is added to the treatment in the therapeutic community. In therapy groups in the therapeutic community 'unfinished business' with parents can be emotionally worked through with the parents only symbolically present (empty chairs, role-played by other residents or by staff members). This can be done in encounter groups (Kooyman and Esseveld, 1984), in psychodrama groups (Yablonsky, 1990), Pesso psychotherapy groups (Jongsma, 1981) or New Identity Process Groups (Casriel, 1980, Maertens, 1986).

In conclusion: the fifteen therapeutic factors of a therapeutic community are all essential for the therapeutic process. If any of these therapeutic principles are not present, the treatment will be less effective.

CHAPTER 4

Treatment goals, tools and techniques

The ultimate goal of treatment in a therapeutic community is not only to enable an addict to live a life independent of drug-taking or other self-destructive addictive behavior. It is not only the disappearance of negative symptom behavior. The goal is also to enhance a positive lifestyle to teach the resident to be able to cope with stress in a constructive way, to change a negative self-concept into a positive one, to learn to sustain fulfilling and intimate relationships with other persons and last but not least, to be able to enjoy life.

The goal of treatment in a therapeutic community goes beyond the treatment of the individual's behavior; this behavior is considered as a symptom of underlying problems. Only after the use of drugs has stopped do the problems appear. In a therapeutic community the individual is taught to regain feelings of trust in himself and in others. The final goal is not neutral, but positive. The resident of a therapeutic community is taught to deal with stress and conflicts in a constructive way. He is also taught that asking for help does not mean being helpless. Therapeutic Community treatment aims at social, intellectual physical and creative development of the resident. In this process it is important that the resident learns to discover the limits of his capacities, to learn that he is not the giant of his dreams, nor the dwarf of his fears and that this is all right (Kooyman, 1986).

The ideal graduates of a therapeutic community program are no longer dependent on drugs and/or alcohol use, they are no longer dependent on the treatment program or any professional help, they do not need to use any mind-altering chemicals and there is no need for any psychiatric treatment or any treatment related to their formal addiction; they are not involved in criminal behavior. They have a positive self-concept and are emotionally open, able to give and receive love, able to handle conflicts and painful experiences, able to ask for help, and to show their needs to other persons. They are self-confident and feel responsible for their own lives. They have developed their social creative, intellectual, physical and sexual abilities. They have learned to be honest, to set limits and to know the limits of their capacities. They function on a social level acceptable to themselves. They have developed a critical attitude towards themselves, the therapeutic program and society. The graduates should be able to function on the same emotional and social level as staff members of the therapeutic program. The graduates are able to seek responsibilities, to engage in rewarding relationships with other persons, to handle and enjoy intimacy and to feel and express emotions (Kooyman, 1983).

This shows that the treatment goal is not only to stop the acting-out behavior, the use of drugs, alcohol or other mind-altering drugs, the use of violence or threat with violence. The treatment goal lies beyond these symptoms. In fact, stopping this acting out behavior is a cardinal rule inside the therapeutic community. The therapeutic community is a drug-free and violence-free environment where the resident can learn to cope with the stress of the life in the therapeutic community without the possibility to act-out this stress in destructive behavior. This has to be learned step by step. The first step is to learn to get used to a regular day and night structure, to assume responsibilities in carrying out various tasks. The resident then has to learn to handle the upcoming emotions in his or her contacts with other members of the community, the fear, the anger and the underlying pain. He or she is confronted with his or her 'image' behavior, the defence machanism to avoid painful emotions. The residents learns to understand his fear of rejection and to change negative self-concepts such as: I have no right to exist, I have no right to be happy, I am not good enough, into positive attitudes: I have the right to exist, to be happy; I am not perfect, but I am good enough. By learning through experience in the structured therapeutic program the resident learns to emotionally work through the traumatic past, to take risks in the present and to acquire a positive view on the future.

The ultimate treatment goal of a drug free therapeutic community is more far reaching than society expects them to be in, to treat addiction problems. However the implicit treatment philosophy is that a person who has been addicted has to change a negative self-concept into a positive one, to learn to cope with stress in a positive way and to be able to enjoy life in order to minimize the danger of a relapse into the self-destructive behavior of the past.

Basic rules

In most therapeutic communities there are only a few basic rules:

- no use of drugs, alcohol or other mind altering substances;
- no use of violence or threats of violence and usually no sex between residents.

Breaking a basic rule by using drugs or alcohol or using violence means immediate expulsion from the therapeutic community. It is important to stress that no discussion is possible on these consequences. The resident may enter again but not without being seriously questioned about his commitments.

During the first months after the start of the Emiliehoeve therapeutic community these rules were not clearly explained to the residents. There was no rule on sex, but alcohol or drugs were not expected to be used and violence was not tolerated. Many times in these first months of the Emiliehoeve these rules were broken. Then long discussions followed usually with the result that the resident who broke the rules would get another chance. After half a year the rules were made clear and the sanction was immediate expulsion. Since then nobody ever used drugs or alcohol in the house which before had happened frequently. This was verified by weekly urine controls on randomly chosen days. In the Emiliehoeve violence was only used by residents on rare occasions. Comberton reported that in the therapeutic community Colemine there has been no act of violence for 13 consecutive years. Clear basic rules are necessary for creating a safe and secure environment (Comberton, 1986).

As to the sex-rule, a rule is necessary to avoid power games and abuse, but prohibition during all phases of the program is counterproductive. Also in this respect the experience of the Emiliehoeve is of interest. At the onset of the program there was no rule on sex. The residents had sex as they had done before admission. As soon as girls were admitted the boys became their protectors. Sex was used to feel good, to manipulate and to get status. When a resident fell in love he or she had sex with another resident to find out how the object of their love would respond. Usually the response was to have sex with somebody else. After about a year, and after long discussions in the group the democratic type of structure was still operative the residents came to the conclusion that sex was abused and could better be banned. There came a 'no sex between residents' until a person graduated from the program. Sex with friends outside was possible during weekends after the first phases of the program. Residents frequently broke the no-sex rule. Sometimes the staff reacted by transferring a resident to another therapeutic community. A resident was rarely expelled from the community. After some years sex between residents was made possible. It had to be preceded by a request. If the residents concerned were considered to be responsible enough and had been in the therapeutic community for a certain period (three to six months) the request was granted. It was no longer acting-out and it could be openly questioned by fellow residents in groups. Since this policy was implemented, the sex rule was rarely broken, while requests for having sex, which could take place in a separate building on the premises, were also rare.

As soon as a resident is in the community, limits are set to acting-out behavior. Drug or alcohol use is not possible and violence is not tolerated. This generates stress dealt with mainly in the encounter groups. Apart from expressing emotions in these groups, there is no other way to find relief from the stress apart from running away. Although a new resident may be well prepared in the induction program, the therapeutic community is a stressful environment and many (15 up to 30%) new residents leave in the first weeks of their stay. The sooner they get used to ventilating their emotions in the encounter groups, the better their prospects are.

The community as an environment to practice direct communication

The therapeutic process in the therapeutic community is based on improving communication. Isolation from contact with other persons is not possible. The communication is schematically improved as shown in Fig. IV.

Figure IV: Communication problems, type of behavior, form of therapy and goal of therapy.

COMMUNICATION	INDIRECT	AVOIDANCE	DIRECT
Behavior	Acting-out	Acting-in; withdrawal; control	Direct emotional response; fight/flight; working through
Therapy (learning)	Limit setting; verbal reprimands; confrontation groups (encounter); creating awareness	Groups on emotions; behavior and attitudes; increasing awareness	Groups; developing new strategies; learning to improve relating with others; increased growth

In this figure one can see how indirect communication shown in acting-out behavior is dealt with by limit setting, verbal reprimands and confrontation in order to create awareness. Avoidance of communication, shown by 'acting-in' behavior, withdrawal and emotional control is dealt with in groups focusing on expressing emotions and changing attitudes. For instance, awareness can be increased in the New Identity Process therapy groups about what is being avoided. Finally oneway communications demonstrated in fight or flight behavior not open to others is dealt with in groups in which a person learns to solve a conflict by a twoway communication using questions and responses. In this way the resident learns to cope with conflicts in a constructive way and the resident learns to work emotionally through interpersonal conflicts with the help of other group members. In this way he improves his relations with other people. Since in a community there is immediate response to acting-out behavior, severe destructive behavior such as automutilation, suicide attempts, destruction of furniture, rarely occurs. Setting limits to acting-out behavior, combined with group sessions in which emotions can be ventilated and also directed towards persons that provoked the emotions, offers the opportunity to develop strategies to cope with conflicts instead of avoiding them.

The therapeutic triangle and circle

The Synanon-Daytop type of therapeutic community offers on one hand a hierarchical structure of resident and staff positions in which the daily activities take place and on the other hand groups in which everybody is equal, in principle, irrespective of the position the person may have outside of the group situation. In this way the therapeutic communities are different from hierarchically organized institutions, such as prisons or armies; the groups offer an opportunity for everyone to confront everyone else's behavior. There is a balance between the circle of the encounter group where everybody has an equal position and the triangle of the hierarchical structure in which the residents have positions with different levels of importance (Dederich, 1975; Kooyman, 1983).

The following elements, tools and techniques are applied in most therapeutic communities for addicts:

The hierarchical work structure

The residents are divided into groups having different tasks such as cleaning the house, preparing the food in the kitchen, taking care of the administration, looking after the garden and the animals, fixing and repairing objects in the house. Each group has a leader or department head, sometimes with an assistant. The others are the crew. There is the kitchen-department, the household-department, the garden-department, the admini-stration-department, even sometimes the research-department and so on. Residents who have behaved very negatively are sometimes isolated from the others and have to do unpleasant jobs like washing dishes. The department-head is accountable to the coordinator; a resident (often more than one), responsible for the daily functioning of the therapeutic com-munity. The coordinator is helped by persons that oversee what is happening and who report on the behavior of the members, called 'expeditors'.

The resident's position in the structure is decided on by the clinical staff of the community. New residents are given positions with few responsibilities. They learn to assume more responsibilities as they go through the process. They can be moved up and down by regular staff decisions. They are often given positions where they can learn by making mistakes. The system offers possibilities to reward persons by moving them to positions with special privileges. They can be fired and sent down the hierarchy to a crew position, not only when their work is not acceptable, but also when they do not show enough concern and interest in people working for them in their department. In this way a resident learns to assume responsibilities and also to delegate them and to deal with authority conflicts. A person having problems with authorities may find himself to be a very authoritarian departmenthead. The resident also learns in this structure to deal with failure and success. Before he leaves the resident is often given some time off, no work position, having time to relax and to think about his plans for the future.

The hierarchical structure can be used to reward positive behavior and put sanctions on negative behavior. Mobility up and down the structure is a tool to reward demonstrated responsibility or to take away a job function with the additional privileges as a response to destructive behavior. In Synanon members were not paid directly and there were unlimited positions to be created. In the therapeutic communities that followed there were not enough places for everyone who behaved well. Not every resident could become a paid staff member.

Rewards and sanctions

Apart from using the hierarchical structure, rewards can be given for positive behavior in the form of certain privileges. One of the dangers in a therapeutic community is that sanctions and punishments may be given much more often than rewards. Taking away privileges or giving unpleasant work assignments, sometimes with bans to speak, can easily be abused.

Limit setting to negative behavior

There are various ways in which the resident is made aware of the impact of his behavior. The most direct ways of creating this awareness are pull-up's and haircuts (verbal reprimands). By having immediate response to negative behavior, limit testing by residents is manageable: they act out by coming late or not doing their job instead of smashing windows or using drugs.

The pull-up

A pull-up is a remark or sometimes behavior made openly, usually upon neglecting something that should have been done. A person is told for instance: „This is a pull-up: You have not yet made your bed". The other person is supposed to listen and to answer with: „Thanks". And he is expected to do what is expected of him. A pull-up can be given to the whole community, especially when it is not clear who is responsible for something negative. Not only the person receiving the pull-up may benefit. Also the one who gives the pull-up reinforces his own positive attitude. If something is not done, usually some mess is found in the community, this can be written on a pull-up list with the name of the one who noted what was wrong. At the daily morning meeting this list is read and the ones who are responsible can make themselves known. This can be used in the meeting to ask another persons' opinion of this behavior and to tell the person what is the supposed attitude behind the negative behavior. Dealing with the pull-up list can be a lively part in the morning meeting and this also gives a possibility to share guilt feelings and get 'clean'.

The haircut

The haircut is a ritualized verbal reprimand. It is applied when negative behavior does not respond to pull-ups or confrontation in the encounter group or when it needs an intense response. The goal is to make a person aware of the destructiveness of his behavior and to show the relationship with his negative behavior in the past. It is also meant to give an indication of what kind of behavior is expected from the resident and in which direction the change of his behavior is supposed to go. The verbal reprimand is given to a resident by at least two, usually three persons who confront the resident with his behavior. The three persons sit in chairs while the person to whom the confrontation is directed has to stand up and is not allowed to respond. He is told to listen and can only respond at the end by saying: „Thank you". This ritual is very effective with residents who always have a response ready in order to defend themselves.

The procedure is as follows: A resident who thinks that another person needs a haircut writes this on a small piece of paper. The form goes to one of the expeditors. The coordinator and the expeditors decide if a haircut should be given after the expeditors have investigated the matter. Then the form of the haircut is decided and a proposal is shown to the staff with a recommendation of the persons who should give the haircut (usually three persons, the head of the residents' department, one of the expeditors and a close peer of the resident). When the recommendation for a haircut seems to be mainly written from personal motives of the person recommending it, the haircut is usually referred to an encounter group. After the decision is taken by the staff upon recommendation by an expeditor or a coordinator that a resident is going to receive a haircut, the resident is told to stop doing whatever he is doing and to go and sit on the haircutbench. This is a seat placed outside the office in which the haircuts are given. It is the same bench on which new residents are placed before their intake interview. He is left there, not allowed to speak, often wondering what may be the reason for the haircut. In the meantime the persons who are giving the haircut prepare the content and the form.

There are many different ways in which haircuts can be given. On most occasions the following procedure is used: in the first 'round' the persons giving the haircut make the resident aware of his negative behavior in the community and express their anger and disapproval of what has happened. In the second round they relate this behavior to his self-destructive behavior in the past. In the third and final round they point out what kind of alternative behavior could be shown in the future reminding the person that he had been able to show other positive changes when he was in the community. Also a learning experience can be given or an instruction to perform a certain task. Thus the message is both confronting and educational at the same time.

Each haircut starts with the words: „This is a haircut, the right to respond has been taken away." The tone of the haircut can be aggressive, spoken with a loud voice. The behavior can be ridiculed. The resident receiving the haircut can be spoken to as an adult or as a child by a parent. A special way of giving a haircut is a 'carum shot' (a term invented in Synanon, derived from billiards) in which the persons giving the haircut talk about the resident but not directly to him. Sometimes a silent haircut is given with no words, or a person is asked to give himself a haircut looking in a mirror. Also haircuts can be given to more than one person. A whole department or peer group can be given a haircut. Sometimes a chain-haircut is given by the staff to the coordinator and subsequently by the coordinator to the department-head and then by the department-heads to the crew-members.

After the haircut has been given and the resident who received the haircut has left the office, the haircut and the attitude shown non-verbally by the resident concerned is discussed by the three persons who gave the haircut. The haircut resident is not supposed to talk immediately to fellow residents about the haircut and should keep his remarks for the next encounter group. Not only the receiver of a haircut benefits but also the persons giving a haircut reinforce their own positive value system. A haircut is not regarded as punishment but a corrective experience. The risk is that they are given too often and therefore lose their impact. Thirty haircuts a day in a community of fifty residents, usually given in the evening, is not an uncommon practice, but hardly effective.

Learning experiences

Another way of making a resident aware of his negative behavior is assigning him to a learning experience. A frequently used learning experience tool of a therapeutic communities was having a resident write a short sentence like: „What am I doing here?", or „I only care about myself" or „I am a thief" on a paper or cardboard sign. The sign has to be worn around the neck and each resident that meets him, has to ask every day why he has been given the sign. Thus the resident has to repeat over and over what is behind that sign. Although this is a very powerful measure, it can also be regarded as humiliating for the resident, especially to visitors of the community who do not understand why it can make sense. Signs can also be abused in having residents walking around with signs for many days. Most therapeutic communities have stopped using signs or changed them from large paper or wooden signs into badges or even small closed envelopes with a sentence inside.

Other learning experiences are to act out parts of your personality or of your defense (your image). The staff can decide to have residents walking around as a nurse, as a patient in pyjamas, as a cowboy or sitting in a corner as a wise man with lots of books, knowing all the answers. Residents can be isolated for some time, for instance told to camp alone in a tent on the premises for a week, when they really want to isolate themselves from other people or can be put in a mock hospital inside the therapeutic community, when they behave as a patient, getting friendly care and attention from fellow residents being mock nurses. These types of learning experiences are opportunities to break the routine and bring humor in the community.

A tough learning experience is sending an older resident who does not think he still has anything to learn, on a research trip for some days, to the places outside where he used to stay before admission. The resident can always come back earlier. The experience lasts often for a few days with a fixed six-hourly phonecontact with a peer in the house. The confrontation with the drug or alcohol scene is usually scary and unpleasant and reinforces the resident's will to put his energy into further treatment.

The encounter group

The word 'encounter' means meeting another person. In the encounter it is possible by meeting the other to learn about oneself. The basic assumptions of the encounter group were already formulated in the Synanon games. In the therapeutic communities these confrontation groups got their name encounter, a name also used for emotional interaction groups developed by professionals in the Human Potential Movement in that same period.

The basic assumptions that underlie the encounter group include:
- Criticism is valuable.
- „Everybody is always wrong"; an encounter group is not a fight about who is right; it is not a discussion group.

–   Nobody is perfect.
–   A group leader is also open to confrontation.
–   If one person tries to tell you something about yourself, it may be his problem. If two persons are telling you the same thing, you should pay attention. When six persons are saying the same thing, you better start accepting what they say.
–   You get out of the group what you put in.

Rogers has called the development of encounter groups the most rapidly spreading social invention of the century and probably the most potent (Rogers, 1970). The encounter group in the therapeutic communities developed separately from the encounter groups of the Human Potential Movement (Schutz, 1975). The latter had originated from T-groups (training groups) set up in the Massachusetts Institute of Technology by co workers of Kurt Lewin (who had been one of the pupils of Moreno, the founder of psychodrama). The encounter groups were more focused on personal growth. In 1967 Schutz started to run encounter groups in Esalen in California, where workshops in new group techniques, such as Gestalt groups, psychodrama, massage, Tai Chi, sensitivity groups and sensory awareness groups were held (Coulson, 1972). In that same period Synanon had many game clubs running where 'square' people were invited to participate. Many of them later joined Synanon to live in the community. Although the encounter groups of Esalen and the Synanon games had a different source they had much in common. Both focused on the here and now experience in the group and both had the goal to create awareness through open communication. In sharing your ideas that are not known to others „hidden agenda's" and „blind spots" become visible.

Already in California in the Sixties the Synanon games and the professionally led encounter groups had influenced each other. In later decades graduates of therapeutic communities joined centers for personal growth as group leaders. In the development of the Emiliehoeve therapeutic community and several other therapeutic communities in Europe these double-experienced ex-addict group leaders taught the professional staff how to run encounter groups in the therapeutic community. Thus the groups in European therapeutic communities often use techniques from other group therapies in order to break through the defenses of the participants. Heavy group attacks on self deceptive attitudes are often the only way to reach the real self of the addicts and that is why encounter groups have become so essential in therapeutic communities (Yablonsky, 1989).

In most therapeutic communities regular encounter groups are held, usually three times a week. One of these three groups is often a weekly special group. This can be a group with peers, persons that have spent about the same amount of time in the program, groups with men or women only, or a specific type of therapy group such as a bonding or a Pesso-motor-group. Regular groups may last one and a half to two hours. It is important to stick to a fixed time, otherwise groups may last longer and longer. Group members may tend to postpone to take part if there is no fixed time to stop. It is important to realize that what has not been solved in a group, will probably be coming up again in a next group.

For an encounter group an optimal size is 10 persons. It should not exceed 12 people, not including the leaders. Too large groups give too much opportunity to hide, not to be actively involved and share in the feedback. In the first period of the Emiliehoeve Therapeutic Community, the tradition was as is still the case in some therapeutic communities to have all residents together in one group (Van Epen, 1990). When the staff had decided to increase the number of residents from 10 to 20, it was noticed that as the encounter group had 14 persons or more, some residents ran away until the group had again stabilized at 12 persons. When the decision was taken to have two encounter groups held at the same time the splitting stopped and the community could grow to 20 and later to 35 residents. The groups were composed each time of different residents. This was another advantage. As the composition of the group was different each time, a problem was confronted differently each time. Instead of a repetition of the same mechanisms over and over again the groups became more dynamic.

Selection of who is participating in which group is decided by the staff and often a proposal is made by the expeditor and/or coordinators. This is done this by examining requests for confrontation written on small pieces of paper by residents. The residents are supposed to „write a slip" when they get emotionally upset by something happening in the community, often negative behavior or the attitude of a fellow resident. These slips are put into the encounter box. By examining the slips the expeditors or coordinators make a proposal indicating who should be with whom in the same group and why.

Although in Synanon the groups were leaderless, in therapeutic communities there are usually two persons responsible for running the group. Two persons have the advantage that when a participant wants to confront a leader, the co-leader can take over and the leader who is confronted is no longer responsible for anybody else but himself. The leaders can share their own emotions to the group or tell how they have had the same experience as what was happening to a group member. In encounter groups self-revelation of a group leader is often practised and seen as an important element of their role. Before the group, the group leaders have a meeting with the staff to discuss what they expect to happen. After the group they come together again to evaluate their group.

In the encounter groups, the persons are confronted one by one by other members of the group (Fig.V). Thus the focus of the group is on one person at a time.

Figure V: The Group Focus.

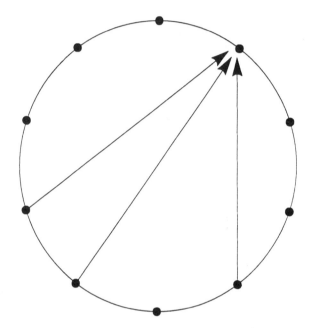

The confrontation can last from a few minutes up to an hour. When a participant does not respond or emotionally freezes when attacked, it is better to stop the confrontation and have the focus on somebody else. Continuing the attack would only have the person reacting more defensively. It usually helps to come back to the person later. Directing the focus at one person is important. Confronting different people at the same time creates confusion. A person is supposed to defend him or herself. Helping a person to defend himself is considered as not really helping as the person does not get the opportunity to stand up for himself. It is like playing the nurse and keeping the confronted person a patient. It is possible that more than one person confronts somebody at the same time. Not what is said is important, but more important is how it is said. This means with emotions. Screaming helps to provoke and ventilates emotions. At the onset of an encounter group there is usually a lot of noise (Blank, Gottgegen and Gottzegen, 1971).

The group usually starts with an angry confrontation by one participant followed by loudly expressed emotions. Other group members join in the confrontation and repressed emotions are ventilated. Usually the attacked person starts reacting often ventilating his own repressed emotions. After the emotional outbursts the group brings in material relating the event of the confrontation to other similar past behavior of the person concerned. Now the confronted person usually stops defending himself, admitting to things that were said. He can experience despair, fear or pain. He is encouraged to express these emotions. The person can be asked to make a commitment to change his behavior. If possible the person is asked to do something right now in the group. The confrontation often ends with a hug from the person who started the confrontation. Then follows a warm phase. There is empathy from the group towards the person that was confronted. A group member can relate that he went through the same thing (Cassuto, 1981). The message is: we still accept you as a person, although we condemn your behavior, we went through the same thing, we are different from you, we will watch that you keep to your commitment. After some more confrontations the group will be finished by asking the group members to share their opinion of what happened in the group. After this feedback the group breaks up to have coffee in a relaxed atmosphere.

Although anything can be said in the encounter group, group-norms do exist (Simon, 1974):

Table 1: Do and do-not norms of the encounter groups.

Do	Do-not
be active, take part	avoid confrontation
make your life public	use violence or threaths of violence
tell the truth	act-out sexually
say what you feel and feel what you say	confront without concern
start with the here and now	play the nurse
seek confrontation	leave the group room
use any language	be silent
polarize and generalize	fall asleep
use contrasts	therapize
show concern	talk outside the group about other
use humor	people's experience
tell your own story	smoke!
use the whole group	
do what you are afraid of doing	
(act as if you are not afraid)	

The norms are simple and easy to explain. In contrast with the leaderless groups of Synanon the encounter groups in therapeutic communities are usually led by two persons.

They can be staff or older residents. These leaders guard the norms. They can also apply some techniques to increase the effectivity of the group. No smoking during the group is important. It can distract the attention from what is happening. Smoking can also be used to suppress tensions and that is opposite to the goal of the encounter group: to express tensions and emotions. The leaders, being role models, should stick to the norms, including abstaining from smoking during the group.

Encounter groups function as a result of specific techniques, mechanisms, roles played by the participants, settings and material. The following techniques are used in encounter groups:

- exaggeration,
- ridicule,
- imitation (of somebody's behavior),
- indirect confrontation by talking about somebody who is present in the group (called 'carum shot' a name derived from billiards).

This last technique is extremely useful with persons who tend to respond to a direct confrontation with an angry denial.

When there is a conflict between two persons they can be asked to confront each other emotionally in the center of the group the so-called dyadic encounter. Use of humor is just as important as ridicule: it is difficult to defend yourself if everybody is laughing at you. The best response is to laugh about yourself. Heavy serious issues in the group should be followed by light or amusing situations. Although the addict situation outside the community was a matter of life and death, the encounter itself is not.

While techniques have to be brought in by the participants, certain mechanisms emerge anyway. During an encounter group the following mechanisms can occur:

- projection: The participant confronts his own problem by confronting the other.
- identification: The participant can often identify with another person's story.
- empathy: Emotional involvement creates a condition of empathy among the participants.
- emotional blocking: A person can emotionally dissociate from what is happening and remains silent for some periods. Further confrontation is useless and has to be postponed.
- transference and counter transference: Emotions felt for persons from the past are transferred to group members. The others may react to those emotions.

It is important to be aware of these mechanisms. They can be referred to in the feedback at the end of the group. Projection is a therapeutically important mechanism. Dederich realized that it is more important what is said by a person himself in the encounter, than what has been said to him. In confronting the other, the person confronts himself (Dederich, 1974, personal communication). It is no use for the groupleader or any other participant to try to hide emotions that arise in an encounter group. The group will notice them anyway, especially from the group leaders. This means that group-leaders should feel free to express their emotions in the group. During the feedback references can be made to emotions expressed in the group that have been transferred from the past.

Apart from the role of group leaders who facilitate the confrontations, there can be a variety of roles emerging among the participants an encounter group:

- The prosecutor: continuously uses indictments to which the accused does not admit.
- The follower: never starts a confrontation but joins the other persons' indictments.
- The witness: supports with personal evidence the indictment.
- The identifier: explains he has been in a similar situation himself.
- The pastor: preaches how the person should behave.
- The mirror: shows the person confronted how he behaves right now.
- The provoker: tries to make a person angry.
- The therapist: searches explanations for a persons' past history.
- The nurse: defends a person when confronted.
- The peacemaker: arranges that the confronted person can become accepted again by the group; suggests agreements to change the behavior in the future.
- The attention diverter: diverts the attention from what is going on by changing the subject, or by childish behavior.
- The creator of insight: explains what is happening in the group.
- The scapegoat: always succeeds in getting confrontations in the group.
- The attention seeker: tries to get the attention by interrupting what is going on.

The playing of such roles is profitable for the player as well as for other persons. When a person frequently plays the same role, this should be confronted as avoidancebehavior. By trying to play different roles the learning possibilities increase.

Encounters can be held in any room. In large communities more than ten encounter groups can take place at the same time. Staff members may walk from group to group through the building asking in every group if there is anybody who wants to confront him and to see if a group leader needs assistance. The persons can be seated in chairs. Through the influence of professional group therapists in European therapeutic communities, the participants usually sit on cushions. Sitting on cushions invites people to touch or to hug each other. The fear of violent acting-out is the main reason why residents in most American therapeutic communities are supposed to stay seated in their chairs. When a lot of anger comes up, the participant may be invited to scream it out, while beating a cushion with their hands or a stick.

In the encounter conflicts in the therapeutic community are dealt with. The group starts with group members confronting present or recent behavior of one of the participants. Often the emotional response represents more than what has happened here and now. Emotions connected with similar experiences from the past may come out. This can be clarified by the group leader after the actual conflicts have been dealt with. This unfinished business from the past can be dealt with in other groups, in marathon (extended) groups, in New Identity Process groups, in Pessomotor groups or in psychodrama groups.

Often it is thought that the screaming occurring in encounter groups is a risk for the person who screams. The opposite is true. Persons at risk are persons that are not able to scream, who block emotionally when screamed at, and who emotionally dissociate from the group. The best answer is to move the focus away from them and to return later in a quieter part of the group. By screaming the person loses the need to act out emotions by his behavior. A persons who can scream angrily at another person does not need to hit that person anymore. Screaming is a tool to bring a person to his emotions. It is not possible to scream for a long time without an emotion in an encounter group. Screaming without an emotion usually after some time becomes screaming with an emotion. Sometimes a participant might scream in a childish hysterical way to get a sympathetic response, a scream without real emotion. The participants pick up the difference and tell the persons to stop. Like all hysterical behavior that does not have the desired effect, the hysterical screaming stops immediately when angrily told to do so. Hysterical behavior quickly disappears in an encounter group as it has no function. Psychotic reactions during encounter groups are very, very rare. They may occur for the first time in his life, when a person is overwhelmed with warm loving feelings and is not able to cope with this. When given personal attention the person may gradually come out of such a psychotic state (Casriel, 1972). After his visit to Daytop Village, Maslow realized that therapists in general too often consider their clients as weak and not able to stand a strong confrontation (Maslow, 1967). Emotional discharge in encounter groups can be beneficial as a catharsis. However cognitive processes to find out what happened are equally important. This process that can take place during the feedback at the end is an important but often neglected part of the group session.

Other groups

Apart from the encounter groups other therapeutic group sessions are usually held. Special groups for men and women are regularly held on various topics, for instance on sexuality. There can be groups on special themes such as birth, the life cycle, authority, guilt. These are usually extended groups, such as marathon groups, lasting about 48 hours with only a few hours sleep and probes lasting 6 to 8 hours. Probes can be held with a peer group. With the younger peers, past hang-ups from the time before they were admitted to the therapeutic community are the focus. With the middle peers, the theme is their position in the community here and now, and with the older peers, their expectations of the future. Socalled static groups can also be held. These groups provide an opportunity for evaluating in the peer groups the progress made during the last week or month. Elder residents may be given a 'hotseat'probe, in which each participant in turn is confronted for about half an hour by his peers. Pessomotor groups can be used to work emotionally through problems of the past. The participants can work on experiences with their actual parents. Group members can play the roles of their parents or of fantasized ideal parents (Pesso, 1980).

Bonding therapy groups or New Identity Groups developed by Casriel do also serve as a method to solve emotionally problems of the past. Besides that, the participants learn to overcome their fear of intimacy, of physical and emotional closeness, and to change old negative attitudes developed in early childhood into positive ones. While the focus of the encounter groups is more on the behavior, the bonding groups focus more on two other basic human elements: emotions and cognition (Casriel, 1972; Coolen, 1985; Geerlings & de Klerk-Roscam Abbing, 1985).

The bonding process can be represented in an eight-step scheme as depicted in table 2.

In the treatment limits are set to negative behavior (I), when positive behavior within strict rules is required, this leads to (II and III) the experience of negative emotions. These negative emotions are caused by negative thoughts especially negative attitudes towards oneself (IV). During the bonding therapy groups the residents learn to express negative emotions and to change the negative attitudes into positive ones (V) and to experience the pleasure of confirming the positive attitudes in the exercises in the group and the closeness of other persons (VI). These corrective emotional experiences lead to positive behavior, combined with positive emotions (VII) (Maertens, 1982).

Table 2: The bonding process in the therapeutic community.

I	Living by the basic rules of the Therapeutic Community means: Stopping the acting-out behavior No use of drugs and alcohol No isolation No violence or threat of violence
II	When you follow these rules, it produces: Fear Anger
III	Behind this is: Pain
IV	The cause of this is: Negative attitudes such as: „I am not good enough" „I have no right to exist"
V	This can change into what really is the case: I have the right to exist I am not perfect, but I am good enough
VI	Experiencing this, gives: Pleasure
VII	The result is: Going to meet people Seeking responsibility Enjoying intimacy Ability to show emotions

Bonding groups can be held once a week. Residents who have been in the house for less than about three months are able to prepare groups with exercises, such as trust exercises, eyecontact exercises and scream exercises. Because working with emotions out of the past may distract residents too much from the learning process in the therapeutic community hereandnow the bonding groups last not longer than three hours. They can also be held as extended groups or in workshops of two consecutive days.

Meetings of the whole community

When a serious crisis has occured or when a serious offence to the norms is made by one or more residents, a general meeting is held. The residents are seated waiting in silence until the staff enters. The director of the therapeutic community runs this meeting. He shows his concern about what is going on and he calls the residents who have behaved badly in front of the group and he tells them what he thinks of them, followed by other members of the staff. Often residents are invited to follow their example. The general meeting is necessary to provide trust again when things have gone out of hand. The general meeting is usually directed at one or more individuals. The meeting can also be addressed to the attitudes in the whole community, like a haircut given to the whole house. On rare occasions a haircut can be given to one member by all residents of the community, usually after a cardinal rule has been broken. In Synanon this was called a fireplace ritual. Meetings of the total community are also called 'house meetings'. A special session with everybody present is a so-called 'copping session'. In this meeting one or two staff members sit in front of the group, representing parent images. They start with a talk on the risks of keeping guiltfeelings inside and then they invite residents to share feelings of guilt about anything with the community. One by one they come forward to speak. They are not given an immediate reaction. The staff members just listen and encourage people to go on. The session usually ends with the staff members telling the residents to perform a task, for instance cleaning the whole house. Such a meeting is necessary when discipline has been low, and a lot of 'negative contracts' between residents are apparent.

Seminars

Daily seminars are given to 'exercise the brains'. Residents, staff members or persons from outside give an introduction to a topic followed by a discussion. Residents learn to talk in front of a large group. They also learn that there may be more interesting things outside, apart from dope. Residents can be asked to tell the story of their life or to talk about a subject that interests them. The philosophy of the therapeutic community or themes such as trust, friendship, prejudices or values are discussed by inviting everybody to give his opinion. A philosophical thought can be put on the blackboard and the groupmembers are invited to comment on what the thought means. Discussions on topics for which there is no solution no right and no wrong help the residents to change rigid thinking and to express themselves freely.

Physical exercises

Daily physical exercises are given early in the morning before breakfast ranging from running around the house to aerobic exercises to music. Sport events are impotant to exercise both body and mind. Sometimes dynamic meditations take place, exercises of the mind and body. These are meditations to music, connecting the participant with his emotions as he follows the music with movements of the body. These kind of meditations were originally used in India.

The morning meeting

Each morning after breakfast everybody comes together in a meeting run by one of the expeditors. It is an opportunity to 'pull up' certain individuals or the community as a whole. It sets the tone for the day and tries to get everybody in a good mood. Most residents need this as they have problems getting a good feeling early in the morning.

The morning meeting starts with an announcement of events occurring that day. Then one of the newer residents is asked to read the philosophy of the community. Another resident is assigned to bring the news from inside or outside the community. After this, the pull-up list is read by the resident who runs the meeting, usually one of the expeditors. The responsible persons are asked to stand up and people are invited to confront the behavior.

The person running the meeting can comment on it or ask the opinion of other residents who are keeping silent. One of the residents reads the saying of the day. This is followed by entertainment, an opportunity to have fun. To close the morning meeting everybody goes outside in a circle with their arms around each other's shoulders to shout the yell of the community. The person having special tasks in the morning meeting can be chosen the previous day. Morning meetings can be very lively and provide fun for the residents and also a lot of information to the staff that may watch the happening.

Other instruments

Every new residentis given a data book to write in what he learns. Staff members have the privilege to read it. When a resident has not been writing for sometime, he will be asked why. The resident can decide for himself if he wants people other than staff members to read his book. From time to time, residents write self-evaluations to be given to the staff before they are interviewed in order to determine if the person is ready to enter a next phase in the program. An extensive selfevaluation paper has to be written before the resident leaves the therapeutic community to go to the reentry program.

Department heads make weekly reports including a progress report on the persons working in their department. The coordinator presents a weekly written report to the staff. New residents who have to become known, can be given an assignment for a few days to call all residents of the community together to come to listen to the news, to what is happening in the community. There may also be a speaker's corner in the community where a resident can speak to the community when he wishes to communicate with everybody.

The techniques in therapeutic communities are numerous. The ones described here are the most commonly used. They are all part of the therapeutic system. Most of them do not need professional training to be able to use them. Only special groups such as New Identity Groups, Pesso-motor Therapy and psychodrama, need the input of trained professionals. For that reason they are rarely used in American therapeutic communities, which have mainly recovered exaddicts on their staff. In therapeutic communities in Europe professional staff members have been available from the beginning of the therapeutic communities, so there is a greater use of special group techniques.

CHAPTER 5

The different stages of the therapeutic community program

The treatment in the therapeutic community is not just a black box in which a resident spends a certain drug free period. It consists of various stages meaningful for the staff as well as for the residents. In most therapeutic communities for addicts, the therapeutic community program is divided into different parts. Before the resident is admitted to the therapeutic community, he or she passes an induction program. Before the resident leaves the program to re-enter in society, there is a re-entry program. The induction and re-entry programs are often situated in separate facilities. The treatment in the therapeutic community itself and in the re-entry program are generally also divided into different parts, called phases.

Although the treatment period in the therapeutic community is the most spectacular part of the program, a good introduction to the program and a well-prepared discharge in the re-entry phase are crucial for the success of the therapeutic community (Kooyman, 1975b). In this chapter the various stages of the treatment are described that exist in most therapeutic communities using the Emiliehoeve program as an example.

The different stages of the program are shown in Figure VI.

The induction program

The induction program prepares a candidate for admission to the therapeutic community. It can vary from one or two interviews to a structured program lasting several weeks or months. The induction programs of the therapeutic communities of the Centro Italiano di Solidarietà in Rome for instance can extend to a period of ten to twelve months (Kooyman, 1987).

Candidates for admission usually have an informative contact at an outreach center, in a detoxification clinic or inside a prison before participating in the induction program. Residents of the Emiliehoeve followed an ambula-

Figure VI: The stages of the therapeutic community treatment program.

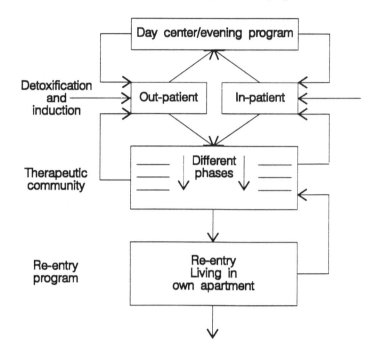

tory induction program up to May 1980, when a clinical detoxification center called 'de Weg' was added to the program. Since then, the induction program is held inside the detoxification center for those who made a choice for further treatment in a therapeutic community or a day center. In the first two years after the Emiliehoeve had been started, the induction phase consisted of orientation interviews at the out-patient clinic for drugs in the city of The Hague. The candidates for admission were seen by the director of the Emiliehoeve in his function as the medical doctor of this out-patient clinic. Further information was given by two elder residents of the therapeutic community who were available in the waiting room once a week. The addict was given the message that the therapeutic community was a drug-free environment. In most cases, during the first 2 years of the Emiliehoeve program the addict was admitted to a department of the psychiatric hospital Bloemendaal before admission to the therapeutic community. Here the candidate was detoxified and visited by staff and residents of the therapeutic community before admission to the therapeutic community.

In January 1974 the induction program became more structured. The addicts could now be seen daily by an ex-addict who had graduated from the program together with two elder residents of the therapeutic community. The candidates for admission were seen in group meetings. The staff's attitude was no longer aimed at trying to get the addict to come to the therapeutic community by convincing him how important it was to get treatment. Instead the staff demanded several tasks from the addicts to show the addicts that they wanted to do something about their problem. The message became: „we do not need you, but you need us." Surprisingly by changing the attitude in this way, more addicts became interested than when they were approached in the traditional way. The addicts were no longer admitted to the psychiatric hospital for detoxification as this environment was not seen as a satisfactory situation to prepare the candidate for the therapeutic community. The message in the induction program was clear: you are responsible for your situation, you can do something to change it but you cannot do it alone. This out-patient induction program lasted seven to ten days. In this period the candidates had to return clean every day. They had to stop using drugs, pills and alcohol. Sometimes they followed a short ambulatory detoxification program at the out-patient drug clinic before participating in the induction groups. Several tasks were given as homework. For instance, the candidates were asked to write down their life stories, to write ten reasons for stopping with the use of drugs and ten reasons for not doing so, to return the next day dressed differently, to read a newspaper and bring up a topic from that paper the next day.

From 1975, inmates of several prisons in the area requested to be transferred for treatment from prison to a therapeutic community. The decision to admit residents to the therapeutic community from prison was taken after many discussions on this issue in the staff meetings. It was feared that the therapeutic community would get an image of being part of the punitive system. Also the motivation of the inmates to seek treatment was thought to be questionable. The decision was influenced by the following incident that occurred in 1974. An addict who had already been in prison several times earlier, had been successful after his arrest in having the police allow him to visit the out-patient drug clinic where he was seen by the Emiliehoeve medical director. The addict entered the room that night with policemen on both sides, asking to be admitted to the therapeutic community instead of the prison. This was impossible but it led to interviews with him in the prison and finally to his transfer to the Emiliehoeve. He got serious jaundice, shortly after his admission, but he recovered and graduated from the program. This incident convinced the staff that their discussion had been academic and that it was merely their problem of how to deal with the new situation, rather than a problem for the addicts in prison.

From 1975, on their own request, addicts waiting for trial could be seen in prison regularly. Those who had been charged with serious crimes such as murder, manslaughter, drug-dealing or armed robbery were excluded from this possibility. When staff concluded that the candidate was suitable for the therapeutic community, the person could be admitted for treatment and the trial was postponed. Usually the trial was held about four months after admission. When they were still in treatment, there were two possibilities: either the case was dropped or the resident was convicted to a prison sentence equal, to the time he had already spent in prison, on the condition that the treatment was continued. In most cases there was also a probation period of one year or more. In recent years addicts can also be transferred to the treatment program after having been convicted. They can spend the last period of their prison sentence in treatment in the therapeutic community.

In the first years the induction program in prison was carried out by Emiliehoeve staff. Ex-addicts who had been ex-prison inmates themselves were excellent role-models for the addicts in prison. Apart from individual interviews also induction groups were organized in The Hague Youth Prison. Once a week there was also a group meeting for addicts who were interested but who had not yet decided to request treatment. After a few years this work was taken over by social workers of the out-patient drug clinic. It was a decision later regretted by the Emiliehoeve staff. In general, it can be said that very little difference in attitude was found towards the treatment between residents coming from the prison and other residents. The choice to seek treatment was made in both groups by the addict. In both groups outside pressure to make a choice for treatment was apparent. After admission from prison, the resident did not have a position different from the other residents. He could leave the program, but he knew that the judicial authorities would be informed if he did not return within 48 hours.

In January 1976 the induction center was transferred from the out-patient drug clinic to the newly opened day center 'Het Witte Huis', a nine to five therapeutic community, modelled after the Emiliehoeve with the psychiatrist who also lead the Emiliehoeve as its medical director. In the years before 1976, the year in which low-threshold methadone programs had been established on large scale, most persons were able to kick off their habit by coming daily to the ambulatory induction program. The less-restricted prescription of methadone to heroin addicts later created a situation in which addicts were dependent on heroin as well as on methadone. This made it more difficult to just stop using drugs from one day to another. It was not easy, it was even impossible to stop 'cold turkey' without drug treatment to alleviate the abstinence syndrome, for addicts who were also using methadone daily. In October 1976, a crisis-detoxification center was opened in Rotterdam linked with the program of the Essenlaan Therapeutic Community, a sister therapeutic community of the Emiliehoeve. Most of the residents of the Essenlaan Therapeutic Community have been admitted since then through this center, after having completed a clinical induction program. To be admitted to the crisis department of the crisis detoxification center, it was not necessary to express the wish to stop using drugs. Only if the client wishes to do so, does a transfer to the detoxification department follow. In this way, a crisis in the drug use of an addict can be used to make the decision to stop, a possible option (Kooyman, 1981).

In the detoxification department of the center, the client was free to choose to participate in the induction program of the therapeutic community. The induction program was integrated in the activities in the detoxification center. For instance, candidates for admission to the therapeutic community attended induction groups while the other residents were cleaning the house or washing dishes. In the detoxification program seminars with information on treatment programs were given together with seminars on subjects such as friendship, honesty, discrimination and trust. In these seminars the residents of the detoxification center were asked to give their opinion of the topic with the staff, bringing the group to conclusions. Apart from these exercises in using the brain to think, also physical exercises and dynamic meditations were added to stimulate physical and emotional expressions. During the first years of the existence of this center at the Heemraadssingel in Rotterdam, more that 20% of the admitted addicts were transferred for further treatment to a drug-free program: for the period between October 1st 1976 and March 1st 1978. More than 20% of the admissions were discharged to a therapeutic community or day center; one third left and within a year were re-admitted (ten remaining left and did not return within one year) (Kooyman et al., 1979). Table 3 shows an overview of admissions and discharges of referrals to this detoxification center.

Table 3: Referrals from the detoxification center Heemraadssingel to therapeutic communities.

	male	female	total
number of admissions	445	140	585
number discharged to drug-free day center or T.C.	88	38	126
percentage of total discharged to drug-free day center or T.C.	19,8%	27%	21,5%

Several clients were admitted more than once before they decided to apply for the induction program (Kooyman et al., 1979).

In May 1980 a clinical detoxification center called 'De Weg' was opened in The Hague. Since then, almost all clients admitted to the Emiliehoeve have been introduced through this center. Addicts from prison were also introduced through 'De Weg' to the therapeutic community. Heroin addicts were given a short detoxification treatment with methadone. This was seen as non-conflicting with the drug-free philosophy of the Emiliehoeve program. In the Emiliehoeve therapeutic community however, methadone has never been distributed.

In the Emiliehoeve induction program a social worker and an ex-addict staff member who graduated from the Emiliehoeve or another drug-free treatment program, formed the induction team in most periods. They were assisted by residents of the therapeutic community, the day center or re-entry residents. The fact that the addicts who entered the induction program met persons with whom a mutual identification could be possible is an important element in the induction process. It could no longer be denied that it is possible to stop using drugs. The ex-addict working in the induction program was a role model for the inductees.

After a period of 7 to 10 days with meetings every day except in the weekends, the addicts could be regarded as ready for an intake into the drug-free treatment program of the therapeutic community or the day-care program. Nowadays, a visit to the community from the detoxification center 'De Weg' is usually possible before the intake takes place. Whether or not a person is regarded as being ready for an intake, depends largely on the behavior in the induction phase. Only persons with severe psychiatric disturbances of a psychotic nature, mental defects, or individuals with severe brain damage were excluded by the staff. The others who did not reach the therapeutic community were those who changed their mind and decided not to participate or to continue the induction program.

A person who had been addicted for a period less than half a year was generally referred to other treatment or an out-patient clinic, for instance, to a day center program. The main criterium for admission was whether or not the addict could show that he or she wanted to do something to change the situation. For those addicts who did not want to make a choice to stop taking drugs, alcohol and other mind-altering chemicals the program could not do anything. Any reason from the addict for making the choice to stop using drugs was acceptable, including the wish to get out of prison.

The intake

The intake in a therapeutic community can be organized in various ways. In the Emiliehoeve it is usually carried out in the form of an interview of the inductee by a group of staff and residents. The future resident has to convince the group that he or she wants to become part of the community. The staff and residents ask why the inductee wants to stop using drugs or alcohol and why help is sought at the Emiliehoeve. It is important that the inductee can ask for help, sometimes the person who wants to be admitted is asked to scream for help. This is something the addict has never done before in his life. The inability to ask for help was in most cases one of the reasons for starting to use drugs, using it as a painkiller. This scream for help is an emotional investment. When this happens the group can really reach out to the new resident.

In the early years of the Emiliehoeve this intake interview could be very emotional. The residents came directly from the streets. Sometimes a commitment was asked or a sign of commitment for instance in cutting off a piece of their hair. Rarely did the interview end in the candidate being refused admission, usually only if it was not clear if the candidate had really made a choice for admission. In that case the person would be told to think it over and come back in, for instance, a week to explain why he or she wanted admission. At the end of the interview the candidate is assigned to an 'older sister' or 'older brother', a resident who has already been in the program for some months and who is responsible for introducing the new member to the other residents and to the values, norms and customs of the community. In the first years of the Emiliehoeve, the newly accepted residents were thrown into the canal in front of the Emiliehoeve farm as a sort of ritual. After this ritual they were given overalls to wear: the uniform of the new members during the first months of their stay. For practical reasons, and due to uncertain factors -the weather, the swimming capacities of the new resident, ice on the water- this ritual was abandoned.

The different phases in the therapeutic community

As the program in the therapeutic community became more clearly structured, the treatment became divided into three phases:

–   Phase I : the newer residents
–   Phase II : the middle residents and
–   Phase III: the older residents.

The first phase was, in most periods, preceded by an aspirant member phase, an induction period into the community life. These residents were called 'eggs', 'youngkees', 'frogs' or other fancy names. The following phases were named 'kangaroos', 'fanatics' and 'diamonds'. The 'egg'-phase residents were guided by a 'mother' and a 'father' resident, who gave them daily seminars on what was important to know in the community. After about two weeks they entered phase I as a full member. In this phase it was important to unlearn negative and destructive behavior and to learn to get used to a clean and orderly life. In the encounter groups the new residents learned to express themselves emotionally and to confront negative behavior of the other residents. In a 'probe' group – lasting a full day – the residents of this phase looked back on those events in the past they felt ashamed of. An important goal of this group was to be open up about their past life in their peer group and to talk about their guilt feelings.

In the second phase the residents integrate the new values of the community. They assume responsible positions in the structure of the house.

In the group they confront their relationships with other members and the associated emotions for persons from past history. In this phase there is a long probe on who they are, their self-concept and their attitudes in relation to others in the community.

In the third phase the residents increase their contacts with persons outside the therapeutic community. Problems with parents, partners and friends are focused on. This period is a preparation for the re-entry into society. A probe is held on the future; what plans are there for further education, jobs, relationships. At the end of this phase, the residents write a report on their stay in the program. In this period they do not have any particular job in the community for one or two weeks. This phase ends with them leaving the therapeutic community to go to the re-entry program.

The re-entry program

The re-entry program is the part of the program in which the resident re-integrates into society. Re-entering from the therapeutic community into society is for most residents a difficult, often painful experience. Many residents think that they have not changed at all in the therapeutic community when they are faced with the same problems they had before admission. They cannot find a job and they feel alone, inferior and worthless. They find out that you cannot always trust people outside, although these may be respected citizens. They cannot ask for an encounter when they have problems with their boss. They find out that they cannot do it alone. They need help and guidance from their peers and from staff. This assistance is usually offered in the re-entry program of the therapeutic community.

There are in fact two models for a re-entry program. One is a 'living-in re-entry'. The re-entry residents are still living on the premises of the therapeutic community. The other model is a 'living-out re-entry' -the residents are living separated from the therapeutic community in their own re-entry house. In the first model the staff of re-entry and the therapeutic community are usually the same. In the second model there is a separate re-entry staff. Both models have advantages as well as disadvantages.

The advantages of the living-in models are:

– The problems of living out and working out are not faced at the same time.
– The residents do not have to get used to a different staff.
– Residents of the therapeutic community have close contact with the re-entry people and can learn from their experiences.

The disadvantages are:

–   The re-entry residents have little opportunity to act out negatively when they come home after a hard day of working in the program.
–   If they have a job as staff-aide in the therapeutic community where they are also living, they will have very little outside exposure. Their position as staff-member as well as re-entry resident is difficult when they are living in the same house.
–   The therapeutic community staff is unable to make the necessary role-shift from parent figure to teacher and counsellor, and may treat the re-entry residents in a similar way to the immature residents in the community.

The advantages of a living-out re-entry program are:

–   There is a clear break with the therapeutic community sub-culture.
–   Inviting friends from outside is usually easier, as well as accommodating partners in the house.
–   The re-entry staff is no longer the authority figure of the therapeutic community and is dealing with the residents on an adult level.
–   Abuse of alcohol or drugs can be dealt with in a different way compared to a situation in which the residents are living in the therapeutic community, where the only possible consequence usually is to throw the resident out of the house.

The disadvantages are:

–   The change from living in the community to living out, coincides with a change of staff and usually change of work.
–   Being a junior resident all over again when entering the re-entry house and having to find your own place.

In most programs however, the location of the therapeutic community has been the most important factor in deciding to have a living-in or a living-out re-entry. Because a therapeutic community is often situated outside the city environment (as is the case of the Emiliehoeve program), a separate re-entry house has been opened in the city. Therapeutic communities however, can be situated in the city like most facilities of Phoenix House in New York. The re-entry program is then usually situated in the same building. The re-entry program of Emiliehoeve is called 'Maretak' and the residents live together in a house in the city in the first two phases of the re-entry program. In the last phase they no longer live in the re-entry house, but in their own apartments outside.

The re-entry residents face a conflict in culture when they leave the therapeutic community environment. In the community the residents learn to be honest and that it pays off to be so. In re-entry they find out that this is not always the case. They may not get the job they want, if they are honest about their past. They are taught in the therapeutic community to confront someone's behavior when it is not okay. They learn in re-entry that people do not always accept their open and honest criticism. They have to learn to solve conflicts without the screaming and yelling they used to do in encounter groups. Their self-esteem shrinks when they have to go through the frustrations of rejection over and over again, while they are looking for a job. It does not impress people when they tell them that they were head of the kitchen in the therapeutic community. They have no degree, no certificates.

Very often the new re-entry resident decides that what he learned in the therapeutic community has little value in relationship with the realities of society. Before he realizes it he is going back to his old behavior. He may no longer tell the truth, no longer clean his room, no longer confront his peers and he may arrive at a point where he is very near using drugs again. It is important to know that this is a common experience in re-entry. Growing a moustache and a beard again can be great fun but wearing the ornaments and outfit of the junky scene can be a sign of being very near a relapse in drug use. The new re-entry resident has to be taught that the values of the therapeutic community are also valid in his life to come. This is especially true in his relation with friends and relatives. He should be taught that it may be wiser at times to keep his mouth shut; but also that telling lies is something he cannot afford to do, because he does not know how to stop.

The same may be so about having to stop drinking alcohol. Many re-entry residents find out that they do not know how to handle drinking. In the first years of the re-entry program of the Emiliehoeve, the residents were not allowed to drink during the program, as they had been informed about the disastrous effect of the so-called drinking privileges in other programs, It was found, however, that some graduates got serious alcohol problems after they had completed the program. So it was decided to offer those residents who wanted it, an alcohol-learning period of two months, starting in the fourth or fifth month of their stay in the re-entry house. They could drink outside the house and report to the staff of their experience. After two months the period was evaluated by staff and peers. Then it was decided whether the person could drink or not during his further stay in the re-entry program. Former alcoholics were never given an alcohol-learning period. They were told to accept that drinking again was too much of a risk. Residents who, before their admission, were addicted to alcohol as well as to drugs were also excluded. It was often difficult for those residents to realize that they had been addicted to alcohol as much as they had been to drugs.

A common problem in re-entry is that most residents have difficulties in leaving behind the safe therapeutic community where you are never alone. Some may even show symptoms of depression shortly after arriving in the re-entry house. Some people have a tendency to act-out shortly before their expected re-entry date, in order to be shot down from their senior position in the therapeutic community and to be able to return to safely washing the dishes again. In this way, their departure from the therapeutic community is postponed. Another common phenomenon in re-entry is the behavior of the adolescent wanting to go his own way, not respecting program values or norms. Although one hopes that the resident works through his authority conflict with the program norms and values when he is still in treatment in the therapeutic community, he rarely does so. Part of growing up is testing out limits and testing must be allowed for, however, without having the resident returning to the former self-destructive behavior. Individual counselling as well as peer confrontation is important in guiding this process.

To make it obvious that the re-entry resident has to become fully independent, the re-entry program was fixed at a maximum of twelve months, at the end of the seventies. The total program was divided into different phases of approximately four months each. In phase I the resident usually works as a staff-aide in the program. Occasionally he can work in a job position outside the program without being paid, still receiving his social allowances. He has a weekly phase I group apart from the weekly evening encounter with the other re-entry residents and a weekly individual meeting with a staff member. These individual meetings are supervised by a psychiatrist. While individual counselling in the therapeutic community has been found of little use, it is regarded as useful in the re-entry stage of the program.

In phase II the resident works outside the program for some weeks without being paid, a volunteer as (a so-called 'trainee') in any job position . Sometimes such a period may have been applied already when he was still a resident in the therapeutic community. After the 'trainee'-period he tries to find a job. He may have his alcohol-learning period during this phase. In the last month he moves out of the re-entry house into an apartment of his own, usually with one or more peers. In phase II he attends the evening encounter once a week as well as a weekly individual meeting with a staff member. In phase III the resident lives outside the re-entry house. He attends the evening encounter once a month and sees a staff member weekly. In the last month before his graduation he doesn't have to come to groups or have to meet the staff. This becomes an option rather than an obligation.

Over the years, it has been found that the change from living in the therapeutic community to staying in the re-entry program was a real cultural shock. To make the transition easier the following changes in the therapeutic community program were made:

–   When a resident of the older peer group becomes a re-entry candidate, he is given a week off to write an evaluation of his stay in the therapeutic community. He also has to find a club outside, for sport or hobby, for one evening a week.

–   He visits the re-entry phase I group once a week. On one evening a week he also visits the education program of the phase I re-entry residents.

As sexual contact is possible among residents of the Emiliehoeve therapeutic community (if the request is accepted) or with persons from outside the therapeutic community, many sexual problems have been already dealt with before re-entry. There are no restrictions as to sex in re-entry. Sexual problems, making new friends, learning to live without the control of the peer group, learning how to spend leisure time, are all subjects that cannot be dealt with sufficiently in the therapeutic community and which make a re-entry program necessary.

As the resident is about to leave the program, former separation fears may be actualized. For many residents, individual counselling is an important part of the re-entry program. In some therapeutic communities (i.e. Hoog Hullen and Breegweestee in the north of the Netherlands) psychotherapy is part of the re-entry program (Vos, 1984). In most programs the re-entry program ends with a graduation party. Some programs graduate clients one year after they have completed the program.

Graduation

In the Emiliehoeve program the client passes the graduation at the end of the re-entry program. In the first decade of the program, the graduation party was an important ritual. The client had to write a paper on the program, or had to show a product of his creative activity. On the day of the graduation, the candidate is interviewed by the graduation committee, consisting of the directors of the different phases of the program, chaired by the program director. A ceremony at the therapeutic community follows. In this meeting the graduate gets a certificate and offers a personal gift to the program. The graduation ceremony is usually followed by a dinner with relatives, friends and a party. It is one of the rituals in the program with great emotional character. In more recent years the client can choose in which way he wants to graduate.

Sequence of phases

Following the treatment in the therapeutic community program the client has to pass through a sequence of phases. Each phase ends with an interview of the individual by staff and residents. During the stay in the therapeutic community, the residents are usually interviewed by two staff-members and some members of the next phase. After passing a phase there may be a ritual. In the first years of the Emiliehoeve's existence, residents were called 'member' only after passing the first phase. Then they were given an Emiliehoeve T-shirt and a small present from the older residents. When the community grew from 12 to 35 residents, the rituals used in passing to another phase were abolished or changed into a single announcement and a speech from the director in front of all the residents.

Passing the complete program, through the different phases with different goals, gives the resident a feeling of accomplishing something step by step; a feeling the resident has often never had before.

CHAPTER 6

The family of the addict and the therapeutic community

Problems in the family

Is there a characteristic structure of the family of an addict? There has been considerable research published on correlations between specific factors in families and the occurrence of drug addiction among children of those families. The relationship between drug abusers and their parents has received considerable attention in research literature. Both over-involved, overprotective parents, as well as detached, neglectful parenting have been found to be related to drug abuse (Kaufman and Kaufman, 1979; Steffenhagen, 1980; Stanton, 1980). The quality of the relations seems to be important. Kandel (1976) found a correlation between the quality of the parent-adolescent relationship and parental drug use and the continuation of the use of heroin and other hard drugs by the adolescent. There is a high proportion of fathers with a drinking problem. Stanton (1980) found in parents of heroin addicts, an alcohol problem of at least one parent in at least 80% of the cases.

Many addicts have experienced a separation or a sudden death of a parent, mostly the father, before the age of 16. However most addicts in Stanton's research did not come from broken homes (Stanton, 1980). Frequent traumatic experiences are reported in families with addicted children including violence, child molest, incest, suicides, psychiatric admissions, sudden death of an important family member, or separation from the parents (Aron, 1975). Reilly (1976) observed that parents rarely rewarded acceptable behavior of the addict, while giving strong (negative) attention to negative (drug abuse) behavior. Anger was rarely expressed in the families he studied. Alexander and Dibb (1975) also found an inability by the parents to express their criticism directly to their children and that there were unclear boundaries between the generations. Reilly and Alexander and Dibb mainly studied middle-class families. Kaufman (1981), who studied families of different ethnic-background and social status, also found vague boundaries between the generations. There were rarely efforts by the parents to stimulate positive behavior of the addicted children. Finally Klagsburn and Davis (1977) conclude in their review of the literature that there is a lack of boundaries between the generations. There is usually a strong mother-child and a weak father-child relationship.

Family structure of heroin addicts

Contrary to common beliefs, most drug addicts do not have an isolated life in the drug scene without contacts with their family. Vaillant (1966b) has found that 72% of the heroin addicts in his study were, at the age of 22, still living in the home of their parents. At the age of 30, 47% were still living with a female relative. Crawley (1971) found in his study of addicts in England that 62% of the addicts were living in their parents' home. Stanton and Todd (1972) found that 66% of their sample of male heroin addicts were living with their parents or at least once a day, contacted their mother by telephone. Haley (1980) described the problems of the families where an adolescent family member leaves home, and the relationship to drug abuse. Stanton and Todd (1975) pointed out the tendency of the addict to relapse into his old habits of drug abuse and to return home when a crisis in the family had occurred after the addict had stopped using drugs and had successfully started to live his own life. They saw at the base of this a separation fear between the family and the addict. The heroin dependency provides a pseudo individuation (Stanton et al., 1978). Alexander and Dibb (1975) pointed out that the addiction is necessary to maintain a certain equilibrium.

In reviews of the literature dealing with family factors in drug addiction (Rose, Battjes and Leukefeld, 1984; Stanton, 1979; Salmon and Salmon, 1977; Harbin and Maziar, 1975; and Seldin, 1972), a common structure of the family of a male heroin addict is described. There is an over-protective, indulgent and permissive mother with the addict as her favorite child. Fathers of male addicts are reported to be detached, uninvolved, weak or absent. Father-son relationships are described as being quite negative with harsh and inconsistent discipline.

In contrast to males, female addicts seem to be in overt competition with their mothers. They see the mothers as overprotective and authoritarian, while their fathers are reported to be indulgent to them, sexually aggressive and often alcoholic. The probability of incest is much greater than normal with estimates as high as 90% in female heroin addicts (Cuskey et al., 1979; Ellingwood et al., 1966).

Research findings on families of addicts in Europe

Survey research on clients of methadone programs in the Netherlands has found that 67% of the addicts had regular contact with their parents (Sijlbing, 1981); 32% (Bindels, 1981) or 38% (Hubert and Van Steijn, 1986) of the addicts were living in the home of their parents.

Research carried out among 50 families of residents at centers of the Association du Levant in Lausanne (Switzerland), showed that in 66% of the cases there was a so-called matripotent family structure. There was a strong female influence over three generations, the grandmother from mother's side, the mother and a sister of the addict or the addict herself when female. There was an alliance between the grandmother and granddaughter, with the grandmother undermining her daughter and criticizing her marriage and the way she raised her children. In 20% of the cases a family structure was found characterized by a mother who separated from a bad husband after the birth of the future addict, later remarrying a man who was the opposite of the first, presenting himself as an ideal husband and father. The researchers in this study point to the important role played by grandparents, the way members are disqualified and the existence of secrets. They frequently found a brother or a sister as a parentified and ideal child in the family. The families were usually closed off from outside contacts (Ausloos c.s., 1986); (Lanini, 1985).

In a research on families of drug-addicts in Zurich the relationship with the parents -especially with the father- was experienced as being worse than in the control group. There was serious family instability in 55% of the addict families. Contrary to other findings in Zurich addicts were found less frequently living with their parents than the control group (Zimmer-Höfler, 1987).

Cancrini (1985) researched families of drug-addicts in Rome in a project evaluating family therapy He described drug addiction related to the family structure and found four types of addiction: Traumatic drug addiction, drug addiction with an actual neuroses, transitional drug addiction and sociopathic drug addiction.

In the traumatic drug addiction there is a rapid break with the normal pattern of life. The drug protects the individual who is in a state of panic. The addict uses, in a destructive way, all kinds of drugs to seek numbness rather than pleasure. The drug use and its dramatic consequences mask or cover the sense of guilt provoked by the traumatic experience. The family context in this type of addiction is rather varied. In some cases the addict has been the 'model' child. Often the addict has recently gone through the phase of individuation and separation from the family while the new network of relations is inadequate for the need in the stressful circumstances.

The second type is drug addiction with an actual neuroses. There is an intense actual conflict in the family around the drug addict. There is a typical structure of the family: deep involvement of one of the parents (generally of the opposite sex); a peripheral role of the other parent; weakness of boundaries in the family hierarchy; a polarity between the addict as a 'bad' child with another 'good' child; contradictory communications in the family and explosive and violent conflicts.

The third type is transitional drug addiction. This addiction has the following characteristics: an experience of powerful, ecstatic, pleasing effects of the drug relieving previous personal sufferings; repeated depressive states with compulsive addictive behavior; difficulties in linking the drug use to specific events in the life of the addict and long-term risk of relapsing often into alcoholism. The family has the following characteristics: there is a lack of definition in the relationships with often, incongruous messages; the members ignore the meaning of messages of others; the family members try to manipulate therapists and others to strengthen their own positions and there can be a repeated, brief but intense acting-out. Both parents are over-involved with the addiction or private life of their children. The polarity between the children is not that of good/bad but rather of success/failure. Often in this kind of family there is a person, named the 'prestigious' member by Selvini Palazzoli (1978), actively involved in solving the problems of the family. At the moment of detachment of this prestigious member, the identified patient reacts by creating a problem through drug use, which has the effect of fixing things as they are.

The fourth type is sociopathic drug addiction. This kind of addiction is found with persons who act out their psychological conflicts. They typically share the following history: known anti-social behavior prior to their drug addiction rapidly adapting a drug-addict life-style, a deviant attitude, an inability to give and accept love. They speak with detachment of their habits and have an underestimation of the effects of the drug. They are usually children of economically and culturally deprived women, or children of multi-problem families in the ghettos of the big cities. Their maladaption is first evident in difficulties at school and later, in adolescence, in the increasing violence with which they react to rules of a society perceived as hostile. Often they were brought up in institutions. The families can be detached or dramatically disorganized.

Cancrini points out that families of drugs addicts cannot be seen as homogeneous systems. The families described in the literature of family therapy with drug addicts are usually of the second drug addiction (in families with actual neuroses) or third (transitional drug addiction) type. Cancrini's distinction in four types of families was found useful in his research-project on the effect of family therapy with drug addicts.

The family therapy development by Stanton (Stanton and Todd, 1982) combined the structural approach (re-establishing the parent-child 'hierarchy' and restructuring alliances and subsystems during the session (Minuchin, 1974) and the strategic approach (focusing on symptoms with specific strategies of intervention, home-work assignment and avoidance of a power struggle with the family (Haley, 1980). Stanton's approach was most successful in Cancrini's research with families of the addiction type with an actual neurosis. A paradoxical family therapy approach developed by Pallazoli (1978) was most successful in families of the transitional addiction type. In families of the traumatic addiction type, individual ambulatory therapy was found to be usually sufficient.

In addiction of the sociopathic type, conventional therapies were found to be rarely successful. Cancrini considers therapeutic communities as a useful tool to fill the vacuum of social and family relationships in addiction of this fourth type. In therapeutic communities a substitute family system is temporarily provided from which the addict can finally separate in a healthy way (Cancrini et al., 1985, Cingolani, 1986).

The therapeutic community as a substitute family

In the initial stage of the development of therapeutic communities for addicts in the United States, the therapeutic community was seen and referred to as the addict's new family; a substitute for a family which the addicts often never had (De Leon and Beschner, 1976). Many residents came directly from prisons. The first therapeutic community modeled after Synanon's self-help program, Daytop Village, started as a branch of the Probation Department of the New York Supreme Court. It was a program for male drug offenders with felony convictions, on probation with this department (Sugarman, 1974). Most residents had never experienced a concerned supportive family environment. The therapeutic community was a substitute family to them. The members and staff of the therapeutic community were called 'family'. Fellow residents were referred to as brothers and sisters.

As the original family was seen as having a negative influence on the residents, for at least four months all contact between residents and family members was prohibited. It was only in the seventies that therapeutic communities learned that involvement of the families in the treatment of the residents could prevent them from activities sabotaging the efforts of the treatment staff (O'Brien, 1983). Instead of being enemies, the parents of the addicts were considered as a potential source of help and support in the treatment process.

The involvement of the family in the treatment programs of therapeutic communities

Although the American therapeutic communities initially excluded parents from the therapeutic process, they were also the first drug treatment programs that made family involvement an integral part of their activities. After the initial years, the staff at the therapeutic communities became aware that the addict, after discharge, will return to his or her family of origin. If the family does not get any attention, rapid deterioration of the benefits, derived from treatment, could occur. It was also evident that other children in the family assume the role of the addicted child in the family, when the brother or sister was successfully treated.

During the second half of the Sixties, a new philosophy, open to involvement of the family was adopted in most therapeutic communities in America. With the opening of ambulatory outreach centers in 1967, parent involvement became structural in the therapeutic community of Daytop Village in New York. Before, there had been informal meetings for parents of residents of the therapeutic community. In the outreach centers younger addicts were treated in a day program. As the residents stayed at their parents' home at night, it became clear that they had to be involved. Weekly groups for parents became part of the day-care programs. Couples did not attend the same parent group. Although the groups were initially only for parents, sometimes other relatives took part (Lakoff, 1984, Maloney, 1985).

One of the first persons working in therapeutic communities who introduced elements of family therapy in the treatment program was Pauline Kaufmann. She was trained as a family therapist by Minuchin. In the early Seventies she started multi-family groups in the day-care program of Phoenix House in New York. Entire families together with adolescents in treatment were present. These groups could be attended by up to fifty persons from fifteen different families (Kaufman and Kaufmann, 1979). The multi-family groups alternated with bi-weekly parent groups in which the clients were not present. Nowadays, the multi-family groups are held once a week, with parent groups only once a month. In the day-care centers in New York a new policy was formulated that addicts were only acceptable for treatment if at least one parent or a significant other person could be involved in the treatment. Parents were trained to be group leaders. New parents attended educational groups on treatment philosophy and clinical methods in the treatment program. Advanced parents took part in a parent encounter group.

In 1980 the parent association of Daytop Village was renamed the Daytop Village Family Association. There are now special groups for couples, for men or women only, for parents of residents in the therapeutic community and for parents of adolescents. For residents in the last part of the program, the re-entry phase, groups are organized of parents and residents together. They have weekly meetings during a period of 2 months, led by a staff-member, a graduate and a parent group leader (Kalajian, 1979). In 1985, groups were formed for siblings of drug abusers. Also a young women's group was set up for wives, girlfriends and daughters of residents supporting them to overcome their feelings of being powerless and victimized (Tarbell, 1985).

The groups in the United States are less rigid nowadays. Also parents whose children are not (yet) in treatment, can come to groups and husbands and wives can come to the same groups. Some programs have started to involve families even during the orientation phase of the treatment (Gleason, 1983). In the Eighties family therapist and therapeutic community programs which had developed separately, started to cooperate and integrate their efforts. Many therapeutic communities in America and Europe started to apply family therapy techniques. On the other hand, the goal of family therapists working with the families of addicts changed. In the beginning, family therapists promoted family therapy as an alternative to in-patient therapy. Later, in an increasing number of cases, they came to support the family in having the addict admitted to a therapeutic community. In this way the addict is given permission by the family to leave home.

The involvement of the family in the Emiliehoeve program

In the Emiliehoeve, parent groups were started as a regular part of the program in 1974. The aims of parent groups were to end the isolation of the parents, to have parents change, while their children changed; to learn to be honest and to increase their self-esteem (Bos, 1977). After the residents had been in the program for some weeks, the parents were visited at home and invited to participate in the parent groups. About half the parents became involved in the parent groups which were held once every two weeks. Most parents started to attend parent groups in the second or third month after their son or daughter was admitted.

One of the primary goals of the parent groups was to reduce the number of drop-outs from the therapeutic community. Most splittees went to their parents if they left the program. They usually stayed overnight with the parents or partners and soon relapsed into their addiction after they left. The parents were told not to let their children stay overnight and to send them back immediately to the therapeutic community. When the parents did not do so, they were told to leave the parent groups. In the parent groups the participants in the initial stages wanted to talk about the problems they had before their child was admitted. Later in the process, parents were able to focus on other problems; problems they had themselves. The parents were encouraged to show their emotions and resolve their guilt feeling.

In the Emiliehoeve program, other relatives were also involved from 1975. Separate groups for brothers and sisters of the residents were started, run by a staff member and an older resident. There were groups for under and over 20 years of age. The activities of these groups varied from playing games to encounter groups. Separate groups for partners of residents were set up. Since relationships sometimes developed between partners of residents, these groups were later changed into partner relationship groups with residents and their partners both participating in the same group. The parent and family program of the Emiliehoeve became influenced in the Eighties by new insights in family therapy. Impaired mourning, often related to the death of grandparents, making every separation a threat; inconsistent messages; the reincarnation of a deceased family member; the born failure; the child taking the position of a deceased parent; the absence of family hierarchy; these are some of the topics discussed in the parent groups (Van der Meer, 1985, 1986). There are separate groups for parents of residents of the therapeutic community and parents of residents in the re-entry phase. Confrontation sessions of the residents with their parents or partners are held at regular intervals during their admission. There are also days on which parents take part in the therapeutic community program for one day.

The Emiliehoeve was the first treatment program in the Netherlands that involved the parents and later other family members in the treatment. Other therapeutic communities in the Netherlands followed this example (Van Dijk-Karinaka, 1985); Frank and Weesie, 1985). Similar parent programs were developed by therapeutic communities in other countries in Europe, for instance in Belgium (Cafmeijer, 1986) and Ireland (Comberton, 1982). In Italy the Centro Italiano di Solidarietà in Rome had the philosophy of offering help when people knocked at the door. Combined with the problem of a limited number of beds in its therapeutic community, the development of an extensive ambulatory pre-therapeutic community program, called Accoglienza (welcome) resulted. These Accoglienza programs originally had the following characteristics:

- The parents and other relatives were involved from the first contacts of the addict with the program. As the clients were then still at risk, most parents agreed to be involved.
- The residents were told to stay at home and do some of the work there. Parents were again placed in the role of authority, supported by staff and parents working as volunteers in the program.
- The addict had to be escorted daily by a family member to the center in the initial weeks and was not allowed to make phone calls, visit friends or handle money. If this was accomplished (usually after a number of failures), the addict had stopped using drugs in most cases. The addict and the parents then entered separate groups introducing them to the treatment program.

This extensive involvement of the family before admission to the therapeutic community (lasting 4 to 9 months) is most probably the main factor for the extremely low splittee-rate of the therapeutic communities in Rome (approximately 95% of the admissions finish the program (Kooyman, 1987; Ottenberg,

1988). Besides taking part in parent groups, parents are active in various committees. They are active in politics, research, prevention, providing jobs for ex-addicts and receiving visitors. Some parents who have the necessary skill are trained to be leaders of parent groups. They should not be social workers or psychologists or other professional helpers: they are just parents. As a result of positive experiments of organizing meetings of parents and children together in the short term therapeutic community of the Rome program, a parallel family therapy program was started. This therapy was added to the parent groups in 1984. Multi-family group meetings and individual family sessions became an integral part of the program of the Centro Italiano di Solidarietà (Gelormino et al., 1985). More than 20 therapeutic communities have been established in Italy following the Rome model. Staff from therapeutic communities in other countries including The Netherlands have visited Rome and studied the Rome experience. They noticed the low drop-out rates of the therapeutic community and realized, that by contacting the parents in their programs only after several weeks following admission, they never reached the parents of the early splittees. Those residents had already left when home visits were planned. In the therapeutic communities in The Netherlands, this led to the decision to contact parents before admission and to start groups in the induction phase of the therapeutic communities of the Jellinek Centrum, the Emiliehoeve, the Essenlaan and other centers.

Although in the United States, Italy, Spain, Greece, Belgium, Ireland and the Netherlands, parents are involved as an integral part of the therapeutic community program, this is not the case in all countries. In some countries the main reason for lack of involvement of parents is that parents are living far away from the centers, as often is the case in Germany, Sweden and England. Sometimes, however, the reason is resistance from the staff, resulting from unresolved problems with their own parents or over-identification with their rebelling adolescent client.

Parent participation in the treatment of therapeutic communities seems to be an essential element of the program. For the same reason, namely that a family may have a negative influence on the addictive behavior of the resident, families were at first excluded from and later involved in the treatment process of the therapeutic communities. One central hypothesis of the Emiliehoeve follow-up study is that parents' involvement through parent groups has a positive effect on the treatment outcome in a therapeutic community.

CHAPTER 7

The staff in the therapeutic community

Professional and ex-addict staff

In the traditional therapeutic community for addicts developed in the United States, almost all staff members consisted of recovered addicts, who were former residents of a therapeutic community. They had made their way without higher education and credentials to positions of therapeutic and administrative leadership. Staff members were role-models for the residents and residents could be future staff members. The professionals who were engaged in the first therapeutic communities usually functioned in administrative or management positions. They were often mediators between the 'crazy' therapeutic community and the 'straight' society outside.

Lakoff (1978) wrote the following on the position of an ex-addict staff member:

> „The ex-addict has much in common with a therapist. He is often bright, highly motivated and resourceful. His drives, which through therapy have been diverted from drug taking behavior, find useful sublimations in helping others. For this reason, with proper training the ex-addict may become a useful para-professional. In view of the training requirements of the ex-addict therapist it would be useful to outline his shortcoming. Twelve months of therapy is not enough to change the deep-rooted mistrust of society which develops in most people who have lived in the 'drug scene' for many years. There is often a bitterness to the 'straight' professional community, especially doctors and psychiatrists, who they feel have failed to help them as they have helped people with other types of problems. By the time the ex-addict has been made a junior staff member, he has adequately learned to control his impulses

and although his resentment of the professional is quickly repressed it often surfaces in passive resistance to the suggestions given to him and by a tendency not to consult."

A problem in therapeutic communities run by a mainly ex-addict staff (i.e. the therapeutic communities Daytop Village and Phoenix House in New York) is that among the staff there often exists an attitude of „what was good for me, is good for others" resulting in maintaining rather rigid concepts and a resistance to change. In Europe, independent of the developments in the U.S., therapeutic communities for psychiatric patients had been developed by professionals within the health system. They were mainly modelled after the Henderson Clinic in England, described by Maxwell Jones (1953).

In the Jones' model the patient was given an important and active role in the therapeutic process. The roles of staff as well as patients were openly discussed in daily community meetings. The professionals in these communities tried to function on equal level with the patients. Recovered patients however were not included in the staff. This was the model for some of the early therapeutic communities for addicts in Europe. It was also the model chosen by the Emiliehoeve staff in the first months of the program. As described earlier, this led to a chaotic situation with decisions which were rather destructive for the residents and the program. Being less in number and decisions being made according to the one-man-one-vote principle, the staff was unable to set clear limits and to provide a clear structure (Kooyman, 1975a,d). Also the staff at other therapeutic communities discovered that the concept of the democratic therapeutic community modelled after Maxwell Jones did not work well in a therapeutic community for addicts (Schaap, 1977). In the Emiliehoeve the Jones model was changed step by step into a model with a clear structure, a clear distinction of roles and responsibilities of staff and clients. In this way in the Emiliehoeve the Daytop Village/Phoenix House model was adopted to replace the Maxwell Jones model (Kooyman, 1975a,d). Ex-addicts who had been working as staff in Phoenix House, New York and London were hired as consultants to help the staff in this transition.

One of the first changes due to this influence in the Emiliehoeve program was that a clear distinction was established between the role of staff and clients. The main function of the program staff was to do a good job and they were paid to do so. The residents were in the program because they could not live a life outside independent of drugs. They needed the program to change this. It appeared to be possible to have professional 'non-ex-addicts' assuming the same staff positions as those of the American therapeutic communities, where they are held by ex-addicts. These professionals could also serve as role-models for the residents of the therapeutic community by showing how to live a healthy and meaningful life.

New professional staff when hired, spent four weeks as a resident in the therapeutic community followed by one week in the re-entry program. In this way it was possible for them to identify with the resident's position in the community. When sister-programs were started in the Netherlands, the staff found that it was better for new staff members to spend this time in a community different from the one they were later going to work in. This avoided role-confusion and transference feelings towards the staff, experienced in the resident period, which could complicate working with the same people as a colleague afterwards.

To work in a clinical staff position the professionals had to be trained in the techniques used in the therapeutic community. For this it was essential to have personal experience with these techniques as a participant. For this purpose, workshops were organized for staff members run by para-professionals and ex-addicts who had been staff members of the therapeutic communities developed in the United States.

The Emiliehoeve staff was trained to run encounter groups and general house meetings, to give 'hair cuts' and to set up morning meetings. In this process of the first years in the Emiliehoeve, the attitude of the professional staff changed from understanding and providing service to teaching and setting clear limits.

The following example illustrates this change of attitude of the author, who was the psychiatrist and director during the first years of the Emiliehoeve. It is the story of his intervention in the case of Jewish girl, Rebecca, who had been addicted for several years, when he met her for the first time. The author will be referred to as K.

> Rebecca was thin, and hid her bad teeth when she laughed. She had tried to commit suicide more than thirty times in several ways. Some years before her admission to the Emiliehoeve she had been admitted to the general psychiatric admission ward of the hospital where K. was finishing his training as a psychiatrist. She was referred from a general hospital where she had been taken after an overdose, which, soon after admission, had led to a heart-arrest, from which she had recovered. She was tense and irritable. K. tried to calm her down by being kind to her prescribing high doses of tranquillizers. Rebecca however, stayed tense and irritable in the ward, breaking cups, plates and windows, occasionally walking into his office, asking for more sleeping tablets and slamming the door of his room, when K. finally refused to give more. Urine tests on opiates were positive, so Rebecca was transferred to a closed ward. The urine tests remained positive. Much later K. found out that a nurse had been involved in providing the drugs. She finally escaped from the ward, was brought back, escaped again and was brought back again. After stealing money from fellow patients, she was discharged with the message that she was never going to be admitted

again. A few weeks later, she was found unconscious at the entrance of the hospital, admitted by the doctor in duty, who had a hard time explaining to his colleagues why he had admitted her.

Being discharged again, Rebecca was admitted to the methadone maintenance programme for hopeless cases, that K. was running at that time at the out-patient drug clinic in The Hague. Although she was shooting amphetamines into her veins before she came to collect the methadon, she managed to convince the staff that it made sense to keep her in that programme. The only positive change in Rebecca during her methadon maintenance period was getting a denture. A year later in 1972 K. started the Emiliehoeve therapeutic community and changed the methadone program of which he was also in charge from a maintenance into a detoxification program. The majority of the 'hopeless cases' of the maintenance program entered the therapeutic community. Rebecca was one of them.

In those days, in the first months of the program, the staff was still very naive. K. was still very understanding. He led traditional group therapy sessions and illegal drugs were still used on the premises, brought in by the residents. Rebecca was behaving badly in the therapeutic community, and at that time K. was not reacting emotionally. He thought she might improve if the staff paid no attention. But her behavior became worse. In that period, nurses of the neighboring psychiatric hospital Bloemendaal, were even more naive than the staff of the Emiliehoeve. They were on duty during the night and on weekends. In one of those weekends Rebecca had managed to disappear from the premises. She came back bringing opium concealed in a doll. Some hours later, a nurse noticed that Rebecca looked drowsy. She said that she had knocked her head against a pillar and that she possibly had a concussion. The nurse took her to bed. He was very kind to her and gave her some aspirins. During a group session that night suddenly a loud noise was heard. When the staff ran into Rebecca's room, her bed was empty and the window open. Down on the concrete floor of the square, Rebecca was lying, unconscious. She was taken to the general hospital with a concussion. When she came back after having recovered from the injuries, a lot had happened at Emiliehoeve. Three staff members, including the psychiatrist had participated in an encounter marathon training, run by a consultant, a graduate of Phoenix House, New York. Immediately after this experience encounter groups were introduced in the community. The staff now free to react immediately and emotionally to the behavior of the residents. As a result the therapeutic community became drug-free and really therapeutic.

Soon after her arrival back from the hospital, Rebecca cut her wrist with a knife. When K. was called, he was furious and he shouted at her that he was tired of this behavior. If she wanted to die there was no sense in staying at the Emiliehoeve. Here was a place where she could learn to live. This reaction was definitely not what K. had learned during his training as a psychiatrist. The incident was however, immediately closed after K's emotional outburst. The wound was treated and Rebecca went back to her work in the kitchen.

In the next encounter group, Rebecca did not respond to questions. It was decided that the group might as well bury her as she could be considered dead already. The curtains of the group room were closed and the group members covered Rebecca with cushions; Rebecca remained silent and did not react. The group started to talk about her, the participants told each other that it was a pity that they had never known Rebecca, that they didn't know who Rebecca really was. Suddenly Rebecca reacted by screaming loudly and throwing away the cushions, shouting: „I want to live". The group participants came towards her and hugged her. Rebecca started to cry and went on crying for a long time. Since that event she has never tried to commit suicide again. Rebecca left the program half a year later. She did not relapse into her old behavior. She married and found a job in a shop.

Much had changed at Emiliehoeve after encounter groups had been introduced. Through this, the staff's attitudes were changed. Suicidal attempts never occurred again. Long before a resident decides to choose such an extreme action his fellow residents have already noticed that something is wrong. It was learned that one of the main causes for change in people's behavior in a therapeutic community is emotional involvement of other people, especially of the staff. Whether they wanted it or not, staff members become parent-figures of the new family of the resident. So a staff member can have the role of a father or mother in a community. They have to be aware of this transference phenomenon. This, however, does not mean that they should hide their emotional reactions. When they are expressed, this is usually experienced by the residents as the concern they need.

Professionals and para-professionals in the staff

In 1975 the first clients graduated from the Emiliehoeve program. They were eager to work as staff aides even before their graduation. The staff supported their idea to come and work in the staff immediately after finishing the program (this was in the 'post-professional phase' described later in Chapter 10). They were seen as necessary role models especially for the induction program. They were not really selected for the job and for many of them the job was too demanding. Often the staff did not recognize the fact that one of the reasons for the choice to be a staff member was a fear to separate from the program and a fear of failing in a job outside. After some failures that led to leaving in conflict or relapsing into addictive behavior with alcohol abuse, new ex-addict staff members were only hired after they had been functioning well in society for at least a year after their graduation.

With the introduction of graduated ex-addicts as staff members in the therapeutic community, the programs in the Netherlands started to experience similar problems to those that had been experienced in the American programs after more and more professionals were hired to work in the program. Serious conflicts occurred between the professionals and ex-addicts, called para-professionals. In the American programs para-professionals had (as later would also happen in the European therapeutic communities) assumed tasks and responsibilities, that had formerly been the exclusive domain of highly trained and very expensive psychiatrists, psychologists and social workers. The para-professional soon found himself being underpaid and asked for a raise in his salary to the level paid to holders of Master's Degrees and Ph.D.s (Bassin, 1973). This and other issues ended frequently in conflicts between professional and para-professionals.

One of the causes of these conflicts is a negative opinion which the ex-addicts have of professionals and their abilities. This opinion is often based on experiences they had themselves in the past with professionals. The para-professional also often regards the work of professionals often as easier than his own job (Weber, 1957). Frequently the ex-addict needs to prove that he or she is good enough to do the job. There are feelings of inferiority, and fears of being a failure and a high need to receive positive feedback.

On the other hand professionals may fear that the para-professionals can do their work much better and that their professional training has no value for this job. In the case of the Emiliehoeve staff, the first ex-addicts were graduates from the same program as the one in which they were now staff members. The Emiliehoeve was for some time the only therapeutic community of its kind in the Netherlands so there were no ex-addicts to be hired who had not been treated in the Emiliehoeve program. The professional staff had difficulties in abandoning their view of these persons as clients. They had difficulties in accepting their change from clients or patients into colleagues. A para-professional - professional conflict was behind the crisis in the program in 1977. (See Chapter 10 'The closed community phase'.)

Personal convictions of staff

An example of problems which can arise when staff in therapeutic communities express their personal beliefs or convictions can be found in the Emiliehoeve history. To provide the necessary training of both professionals and ex-addicts, a common training program for staff in the Netherlands was started in 1976 in The Hague (Laudermilk, 1981). The first American consultant of the Emiliehoeve program who had been excellent as a teacher and group leader, had been appointed as the director. Shortly before assuming his new position, he had become a member of a sect called 'Sanyassin' headed by a guru called Rajneesh, who had named himself Baghwan (God). Followers were all clothed in orange and wore a necklace with a picture of Baghwan, called mala. Notwithstanding that the new director had promised to the Board of the Training Institute not to speak of his personal beliefs, at the end of the first year of the course more than half of the residents wore orange clothes and became Sanyassins. This resulted in the resignation of the program director of the Emiliehoeve as president of the Board of the Institute. Soon after this the government withdrew its funding from the Training Institute. At the end of the second year this Dutch Training Institute -where professionals and ex-addicts were trained together in a common program to become addiction therapists- had to be closed.

Among the students at this institute had been a large number of staff members of the therapeutic community Essenlaan in Rotterdam (most of them had become Sanyassin). A possible explanation for the fact that the majority of the Essenlaan staff joined the Baghwan-movement compared with only two persons of the Emiliehoeve staff, is that the psychiatrist who had been director of the Emiliehoeve as well as the Essenlaan program left his position in the Essenlaan in 1977 to concentrate all his efforts on the Emiliehoeve program. This left the Essenlaan program without a charismatic father figure. A year after the majority of the Essenlaan staff had become Sanyassins, more than half of the re-entry residents of the program had followed their example and had also joined this movement wearing orange clothes and mala's. As the Sanyassin philosophy conflicted with the program philosophy by being less strict in the attitude towards drug use and sexual norms, the program of the Essenlaan therapeutic community - which had been a copy of the Emiliehoeve therapeutic community until 1978 - changed dramatically. Sex between residents of the community was allowed and even promoted as a good development in the program. Smoking cannabis was no longer forbidden in the re-entry part of the program.

In the Emiliehoeve therapeutic community, two staff members, who had been trained at the Dutch Training Institute, had become Sanyassins. A program policy was adopted, demanding that members of sects, religious or political groups, had to wear ordinary clothes and no visual signs like mala's or buttons showing that they belonged to a certain group. They had to use their official names and had to abstain from talking to residents about their personal opinions with the apparent goal of conversion. After this policy was introduced, one of the orange staff members left, while the other one wore ordinary clothes during work and did not use his Sanyassin-name (given to him by Baghwan). The staff of the Emiliehoeve felt that this policy had been crucial in preventing sects, cults or other personal belief systems from interfering with the program's philosophy which was to help residents to become adult persons, each with their own choices in life, independent of the therapeutic community program. Compared with the Essenlaan therapeutic community only a small group of the Emiliehoeve graduates joined the Sanyassin sect during re-entry or after graduation. The Sanyassins had also been a problem in Germany were staff members of Daytop Germany had been trained by the same charismatic group leader who had been the director of the Dutch Training Institute for addiction therapists.

The above example illustrates the danger of the interference of sects with the goal of therapeutic communities. A direct result of the free sex norms between residents of the Essenlaan therapeutic community and new residents who could not cope with this freedom was a high drop-out rate. Smoking cannabis in re-entry was frequently the first step for residents to resume the use of other drugs. The Essenlaan therapeutic community was for some years regarded as a cult and the program became isolated from the other treatment programs in the Netherlands by the Baghwan's norms of free sex and free drugs. Staff at other therapeutic communities considered that learning to postpone the immediate fulfilment of one's needs was a valuable concept of the therapeutic community program and denounced the obvious indoctrination of residents with norms of a particular sect. After the Essenlaan staff had stopped wearing orange clothes and new staff not belonging to the Sanyassin movement were hired, the cult image of the Essenlaan disappeared.

The 'Sanyassin' period of the Essenlaan started after the period when the psychiatrist and program director of the Emiliehoeve was also director of the Essenlaan. The Essenlaan resident cohort of this follow-up study was treated in the therapeutic community during this joint program directorship and before the 'Sanyassin' period.

The cult phenomenon

The therapeutic community does not require staff to embrace a cult from outside to become a cult in itself. Ottenberg (1984) gives the following description of a therapeutic community:

> „The entire community meets daily, at which time various members speak about their experiences and feelings towards the community. There is a good deal of group singing and group physical activities. Memorable community anniversaries and significant events are noted with special celebrations. There are periodic feasts and festivals.
> Members who have been in the community for a longer time have more privileges than more recent arrivals along with special responsibilities in

the orientation and indoctrination of new members. All members must abide by community rules; those who don't are singled out and penalized. Everyone observes the dress code. All members share in necessary labor and participate in various activities to raise funds for the community.

Members are not permitted to leave the community grounds without permission from the authorities. Trips outside the community are made only in groups: no new member ever goes 'out' without 'support'. As a member of the community one gets to understand that the worthiness and high ethical standards one expects to find in the community, are not present in the outside world which is perceived as dangerous. Prohibitions against communication with outside families and friends is necessary, particularly in the early phase of membership. These restrictions are strictly enforced.

The community creates and uses a private language made up of words with new definitions, phrases, mottos and newly-coined words all of which have special meaning to members and are not readily intelligible to outsiders. One learns secrets known only to the membership. One must have faith in the community. Older members are 'role models' who help newer members to 'trust the leaders' and 'trust the process'. Sometimes group sessions lasting many hours are used as a means of lowering resistance and penetrating psychological defenses.

As a member you experience the joy of being part of something greater than yourself. You can't really understand it unless your experience is personal. And to experience it fully, you must 'let go', 'give up yourself', which means abandoning all questions and doubts and immersing yourself without reservation in the community's activities and beliefs. In the early part of membership, before one's faith in the community and its norms are solid, one is expected to 'act as if', that is, act as if one is fully convinced, even though one's conviction is still tentative. Do what you are expected to do whether you like it or not, understand it or not, accept it or not, or motivated or not. Later you can concern the emotional and intellectual considerations. Right now the only consideration is behavioral: Do it! One's entire life will be different, and better, as a result of membership in the community. One owns full allegiance and loyalty to the community in return."

Ottenberg points to the fact that this description of a therapeutic community fits many of the therapeutic communities and also that it fits many cults, equally well. Ottenberg provides the following definition of a cult:

„By their definition a cult is a group that exhibits the following characteristics:

- It is a group of people who follow a living leader, usually a dominant, paternal male figure, or occasionally, a pair or a family of leaders.
- It is a group whose leader makes absolute claims about his character, abilities, and/or knowledge. These claims may include any or all of the following: a claim that he is divine; God incarnate, the messiah, etc. A claim that he is the sole agent of the divineon earth; God's agent or emissary. A claim that he is omniscient and infallible - the possessor of absolute truth and total wisdom.
- It is a group in which membership is contingent on complete and literal accep-tance of the leader's claims to divinity, infallibility, etc., and acceptance of his teachings, doctrines and dogma.
- It is a group in which membership is contingent on complete, unques-tioning loyaltyand allegiance to the leader.
- It is a group in which membership is contingent on a complete and total willing-ness to obey and the cult leader's commands without question.
- It is then a group that is by definition undemocratic, absolutist.

Not all groups labelled 'cults' show all of these characteristics to the same degree and a group need not be religious in nature to be a cult. One of the implications of the above definition is that it does not assume that any of the more objectionable actions commonly ascribed to cults, are necessarily intrinsic to their nature. To put it another way, the fact that a group is a cult does not necessarily mean that it has to raise funds under false pretences, recruit members through deception, counsel hatred of parents, distort the beliefs of other religions, violate the laws of the state, or forbid its members to receive medical attention. On the other hand, there is nothing intrinsic to a cult that would prevent any or all of these practices from taking place; in a cult everything depends on the leader." (Ottenberg, 1984).

The change of Synanon into a cult

In Synanon, the group, the collective, had become more important than the needs of the individual like in primeval cults, described by Weber (1958). In those cults all individual interests were left out of consideration. God was worshipped to provide good fortune for the collective as a whole, such as rain, sunshine, victories over enemies. For personal problems, sickness and other evils, one turned to magicians, elders, and priests for advice.

Synanon had acquired the status of a religious organization primarily to get a tax exemption as they were no longer regarded as a therapeutic community. The closed community of Synanon had no longer the intention to help their residents to find a place in the wider society. It was a program for life. It had also allowed its founder Dederich to assume unquestioned power. In a journalistic account (in the New York Times of November 27th, 1978) on Synanon, the residents had been described as perpetually smiling former misfits worshipping the founder as if he were a god. The residents, first men only, later women, shaved their heads as a sign of commitment.

When Synanon admitted large groups of juvenile delinquents for a special project, it was found that these youngsters were extremely difficult to handle. For the first time the non-violence rule in the organization (physical violence meant removal from Synanon) was broken. Synanon members were allowed to hit these adolescents, when they thought this was necessary. This was the first break away from the original ideals. The leader and the community fostered suspicious and paranoid ideas. Martial arts and training in defence tactics had been introduced. Armed guards were stationed at Synanon's gates. Allowing one person to become a dominant leader beyond challenge had put the entire community at the will of this person.

> When the author visited Synanon in 1974, Dederich expressed to him his worries of no longer being confronted by members of Synanon on his own behavior. He felt that the residents of Synanon saw him as a demi-god. Even in the Synanon game he was only challenged by his wife Betty, his brother and his daughter. Although he was apparently aware of the dangers of this situation, it did not change into something better.

The situation became worse after the death of Dederich's wife Betty in 1977. Dederich directed all married members to divorce and pair off into three year 'love matches' with new mates. He had also forced all men over eighteen to have a vasectomy. Dederich had decided that Synanon was not a good place for children. In fact, he may have feared the confrontation with persons in Synanon who had not chosen to be there. In a reaction to official criticism of the organization, Synanon declared itself independent from the United States and tried to open an Embassy in Washington. After a lawyer (who had won a lawsuit against Synanon and had been successful in having several young persons removed from Synanon by court order) was almost killed by a rattlesnake put in his mailbox, Dederich was arrested. He was charged with conspiring to commit murder using a rattlesnake. As a piece of evidence a tape was used of a radio transmission on Synanon's own system by which the games of the elders of Synanon were broadcast. When the activities of that lawyer against Synanon had been reported, Dederich had screamed: „kill that man". The next few days two Synanon residents had put the snake into the mailbox. He was convicted to a prison sentence and ordered never again to get involved with Synanon or any organization rehabilitating addicts.

Synanon had broken all of the rules that safeguard a therapeutic community against a charismatic leader who becomes incapable of continuing as the major source of moral and ethical standards (Ottenberg, 1982). Even the Synanon game had ceased to be a corrective instrument. Synanon's leaders rarely participated in conventions or meetings with leaders of other programs. Synanon was proud of remaining self-supporting without help of the government or foundations. Because of this, Synanon avoided any obligations to be accountable to tax-payers or for the community activities (Endore, 1968). What happened to Synanon resembles what happens in many cults when the leader is no longer challenged. Sometimes one wonders if the outrageous behavior of such leaders may be an effort to find boundaries. The decision to have couples change partners and all males sterilized, may have been such an effort. The wish of the Baghwan to own more than ninety Rolls-Royce limousines may have been a similar effort to find boundaries.

What are the safeguards to prevent therapeutic communities from changing into a cult? First of all, the goal of a therapeutic community is the therapy for which the clients are admitted. Everything that occurs in the therapeutic community must be part of this goal. The community is a therapeutic environment; no more and no less. It has to serve the individual who lives in it and not the other way around. The therapeutic community program has to help the individual to function autonomously outside the therapeutic community. Thus, power should never be in the hands of one person. The leaders should be responsible to a group of outsiders, representing the wider community like in most programs where leaders are responsible to a Board. Contrary to a cult, the therapeutic community should focus on the unique identity of each person. The therapeutic community strengthens the ego, the cult suppresses it (Ottenberg, 1982).

Abuse of power

Although most therapeutic communities have safeguards built into their organizations, abuse of power by leaders of therapeutic communities is a potential danger of these programs. In the case of the Emiliehoeve Therapeutic Community as described in Part II, the temporarily appointed foreign ex-addict director didn't allow himself to be challenged by his junior staff and by the residents. He was also not sufficiently challenged and controlled by his supervisor and by relative outsiders. Many examples of therapeutic programs with a charismatic figure-head exist. One example is the program 'Le Patriarch' in France founded by Engelmayer who is its charismatic leader. This program is regarded with suspicion by professionals in the field. Sometimes these critics overreact. This overreaction unfortunately only helps to keep these programs even more isolated than they already are.

Abuse of power occurs when a staff member has a need to be powerful, often as a result of his own feelings of insecurity. These feelings of insecurity can be overcome to a large extent by an extensive training program. This training for people working in therapeutic communities is necessary for ex-addict staff as well as for professionals (Vamos & Devlin, 1975; Ottenberg, 1986). Professionals who lack the experience of having lived as a resident in a therapeutic community need to have been in such a position for some time to fully understand its meaning. Professionals who were selected to work as clinical staff members in the Emiliehoeve Therapeutic Community had to be a resident in a therapeutic community for one month. Preferably this therapeutic community was a sister community to avoid role-confusion. Ex-addict staff were obliged to follow a part-time training course at on an Academy for Social SciencesIn the short period that the Training Institute for Addiction Therapists existed in The Hague, professionals and ex-addicts followed the same training program. In this way they learned to appreciate their own and each other's skills.

Staff burn out

But good training cannot prevent all work problems. Unfortunately, a common phenomenon among staff in therapeutic communities as well as in other help institutions is the staff burn-out syndrome (Majoor, 1986). It is characterized by a person wearing out or becoming exhausted by excessive demands on energy, strength or resources. It usually occurs about one year after someone has begun to work. The physical signs are a feeling of fatigue, physical complaints, headaches, gastro-intestinal disturbances, sleeplessness, loss of weight and shortness of breath (Freudenberger, 1976, 1980). It is found among over-committed persons often with a sub-satisfactory social life. They usually have extremely high aspirations and are setting themselves up to failure. Other persons proven to burn-out are individuals who need so much to be in control, that no one else is regarded as doing the job as well as they can. They have to do everything themselves. The administrators of a program run a special risk. They may become the only person to write a proposal, meet a reporter, confer with the mayor, handle staff training and approach foundations for money. They also have to switch constantly from the language and culture of the therapeutic community to the culture outside. As Freudenberger states:

> „For instance to use the word 'fuck' in a training session is expected. But to utter such a word in a TV-interview is not cool to say the least. In order to guard himself against such a blunder (if such words are part of his usual, easy way of speaking) requires a tremendous effort and concentration. The burned-out person's behavior changes. He becomes easily irritated and develops a paranoia. The paranoid-like state may be

increased by feelings of impotence. The staff member may take risks too easily in his work. The behavior may even be of such a nature that being fired is the result, for instance, by using drugs or having sex with a resident. Another manifestation is becoming rigid and closed to any input. Change is threatening to an exhausted person. Further indicators are a developing cynical attitude, hanging around, spending more time at the job, doing less".

According to Freudenberger sending already burned-out staff to a group therapy session is not such a good idea. The person needs sympathy and support. The group session may just add to the stress.

The following prevention measures can be taken to intervene against the burn-out syndrome:

1. A good selection procedure.
2. Rotating functions; staff do not have to do the same work over and over again.
3. Teaching staff to delegate responsibilities.
4. Having staff witth normal working hours, utilizing their free time.
5. Sufficient time for holidays.
6. Sending staff to training workshops and conferences.
7. Keeping the staff in a good physical condition by encouraging involvement in exercises, making them physically tired.

Volunteers

One should be careful with volunteers. Before involving them in a program, it is important to verify their motives, especially if they already have a stressful paid job. Working with addicts can attract people who need the work to forget the sorrows of their own dreadful personal life. Addicts are persons who are definitely more miserable than they are. Working with miserable people makes you forget your own misery. This, however, may prevent real change. Such volunteers may keep the person they 'help' unconsciously in the same position. A healthy person may confront them with their own misery. This phenomenon is not only present among volunteers; paid staff may also have this unconscious need to be involved with persons in relation to whom they are superior. Volunteers, however, as they are not paid, are more likely to have the satisfaction of this need as an unconscious drive. On the other hand volunteers can be useful for the program especially as representatives of the wider society. They should always pass a selection procedure and if possible a special training program.

The staff structure in a therapeutic community

In the first ten years of the existence of the Emiliehoeve Therapeutic Community the staff structure had greatly changed. In the first half year the staff was structured horizontally. Decisions were taken in the staff by the whole group. That group was responsible which meant that no person alone was responsible. The psychiatrist who had founded the Therapeutic Community was one of the group members. However, the outside society and the hospital organization to which the Emiliehoeve belonged, regarded the psychiatrist as the responsible person. In fact, he was responsible for the treatment according to the law. Realizing this, the psychiatrist reluctantly took the position of the leader of the group. It was only in September 1973 (one and a half years after the program started) that the staff was structured hierarchically with the psychiatrist as the director, the psychologist and his wife both as assistant directors (leading the clinical staff and staff with special tasks), the social worker, the supervisor of the farm, the creative therapist and the secretary. In 1974 one of the clinical staff members was given the special task of setting up a program for parents of the residents. In 1976 when the program had grown to a Therapeutic Community with 45 residents and about 20 persons in various staff functions, the staff structure of the whole program was as follows (Figure VII):

Figure VII: The staff structure of the Emiliehoeve Therapeutic Program.

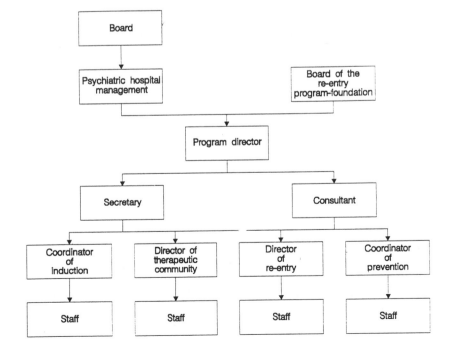

The program director was the psychiatrist who was ultimately responsible for the treatment program. He was accountable to the management of the psychiatric hospital of which organization the Emiliehoeve was a part, and to the board of the re-entry foundation responsible for the re-entry program and accommodation. The parent program was run by the staff of the prevention unit. The program director was also head of a separate drug-free day program called 'Het Witte Huis' in The Hague. This was a nine to five program modelled after the Emiliehoeve organization within a different treatment organization. The program director and the directors and coordinators of the different units met weekly for a staff encounter followed by a business meeting.

The staff structure of the therapeutic community was as follows (Figure VIII):

Figure VIII: The staff structure of the therapeutic community.

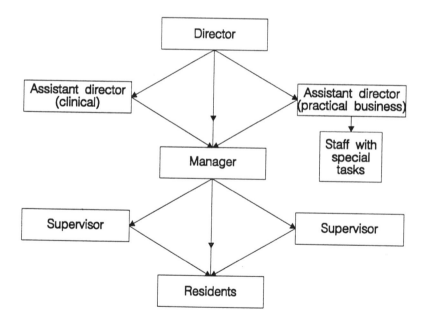

The residents had direct contact through their hierarchical structure with one of the two supervisors or with the manager. The assistant directors could replace each other. However, as a special task one of them had the organization of practical business and guided staff with special tasks such as: the creative therapist, the farm supervisor and the teacher. For a short period there was also a supervisor for the work of the residents' maintenance department. The other assistant director had special tasks in the therapeutic activities, such as planning and running special groups, organizing resident evaluations and clinical meetings. The director was responsible for the overall planning and guidance of the staff. The manager directed the day-to-day activities with the supervisor assisting him and gathering information on what was happening in the community. This structure was modelled after for the hierarchical residents structure. It was largely copied from the staff structure in the American therapeutic communities. A psychiatrist in training had a special position in the staff. He was accountable for his daily work to the assistant director with special therapeutic tasks but he also had a direct line with the psychiatrist-program director on medical matters.

After the temporary foreign director had left, a graduate of the Emiliehoeve took over as the director of the therapeutic community. He was a good therapist but lacked management qualities. After a year the assistant director who had been responsible for organizing practical matters in the therapeutic community was made co-director. This was in line with a trend to introduce dual management in health institutions, dividing therapeutic and organisational responsibilities. The director was responsible for the therapy, the co-director for the organizational management. Although this may work in psychiatric hospitals it did not work in the Emiliehoeve. It is not a good model for a therapeutic community, because unlike in psychiatric hospitals, all activities in a therapeutic community are part of the therapy: the groups, the cleaning, the meals, the cooking, the maintenance of the buildings. The residents are responsible for running their house as part of the therapy. After some years of try-out, one director of the Emiliehoeve Therapeutic Community was appointed when the previous director resigned and the co-director became the director of the re-entry program. The lesson was learned that a director of a therapeutic community should not only be a good therapist or group-leader but that he should also have good administrative and organizational skills. The latter are of more importance than the former. Having administrative and organizational skills is a must for all directors or leading positions. This was also shown above in the description of the training institute, where the director was chosen mainly on his therapeutic qualities. Not taking into consideration the need for administrative talents is a mistake which can became a disaster.

Staff encounters

In the Emiliehoeve program all staff meets weekly for an encounter group, usually before the business meeting. If this meeting is not scheduled weekly conflicts are left unsolved. It has been found useful to have an outsider, (e.g. a consultant), leading these staff encounters from time to time. Otherwise the director will usually lead the encounter. This prevents working out conflicts between the staff and the director. Often, only aggressive emotional attacks are directed at the leader without a solution being found.Staff encounters can prevent irrational attitudes developing among staff toward each other. It may prevent some problems and may solve others. Not all problems can, however, be prevented or solved by encounters.

Meetings of the entire staff

When the program of the Emiliehoeve grew and staff started to work in different units, it became necessary to have a meeting of all groups together. From 1976, meetings were held several times a year with all staff including the staff of the sister program, the day-center 'Het Witte Huis'. These meetings were initially a reaction to the crisis with the director of the therapeutic community, who had isolated himself from the rest of the program. Its purpose was to avoid splits between the different units of the program in The Hague. It was also a means of creating a unity in the drug-free treatment in The Hague. Although there was one program director, the activities were the responsibilities of three different organizations: the psychiatric hospital 'Bloemendaal', (of which the detoxification center 'De Weg' the therapeutic community 'Emiliehoeve' and the prevention program were a part), the addiction treatment and consultation center Zeestraat (to which belonged the day center 'Het Witte Huis' and the ambulatory induction center) and the foundation of the re-entry program 'Maretak'. In one of the first meetings a council was formed consisting of all staff members of the different units. Apart from discussions on program policies for staff (i.e. the policy on membership of sects), staff gatherings with an informal character were also scheduled. Once a year there was an experiental workshop led by an outsider and once a year the staff went on an outward-bound trip climbing mountains, canoeing and camping for a few days. This meeting had a good preventive effect keeping the staff united. A split in the staff is disastrous in a program. It immediately has negative effects on the residents.

In a Swedish follow-up study, a clear correlation was found between a low success score among residents and leaving the program during a staff crisis (Franeer, personal communication 1976). Staff stability is of great importance to a program. „Residents cannot grow further than the staff's level" is a saying among staff in therapeutic communities. Good training possibilities for job changes, possibilities for consultation and mutual confrontation, can all help to stabilize the staff situation.

In the course of time in the Emiliehoeve program, there have been some negative and some positive experiences in the staff-formation. Of the positive experiences the following can be mentioned:

1. A great help was the appointment of outside consultants. They gave advice to the staff through the director in the first year and later through the program director. They were ex-addict staff from therapeutic communities in the U.S.A. or England. Especially in the first years they were of great value in developing the program. Their recommendations could be followed or not, leaving the responsibility in the hands of the director of the program.
2. Different from most American programs, professionals were hired who did notbelong to the clinical staff structure. They were not responsible for the therapy in a strict sense but for the supervision of the work in the resident departments (the farm, the maintenance and the creative departments) or in case of the teacher, for classes in basic education and remedial teaching to individual residents. They were not involved in the therapeutic and disciplinary actions of the staff, but served as representatives from the society.
3. Having assistant psychiatrists working in the therapeutic community for one year during their training, was an opportunity to have professional trainees with a knowledge of therapeutic communities. They were useful for maintaining good relationships with professionals.
4. Having a psychiatrist as program director was, especially in the early years, important to protect the program from being seen as non-scientific and unprofessional.

Some of the negative experiences were:

1. Having a foreigner as director created unnecessary communication problems, especially because he did not understand the Dutch language. It brought the therapeutic community in a state of isolation from the wide society and increased the pre-existing paranoid tendency.
2. Having a husband and wife working in the same staff, which occurred twice, was complicating. They were seen as one unit by the staff and residents when they confronted each other in meetings. It caused almost a crisis in their relationship. When they confronted a resident it was felt as if the spouse was seeing them in the same way.
3. Having minority groups represented in the staff is excellent and especially necessary for minority group residents. However they should preferably not be in a majority position in the staff (i.e. no majority of blacks or homosexuals) as the residents need to have a representation of society's reality reflected in the staff.
4. Graduates in the staff are important role-models for the residents.

However, when they become staff in the program where they were residents themselves it is difficult to get away from the former client-staff relationship with the other staff members. It is therefore preferable to recruit graduates from other programs as staff in the therapeutic community.

5. Graduates in the staff who had not had work experience outside of the program and who had not had an education became too much dependent on their job inside the program. No one expects them to work in the therapeutic community until they are sixty years old so it is important to provide them with the possibilities for alternative choices.

CHAPTER 8

The limits of the therapeutic community

Negative and positive motivations

In the treatment of addicts we have to deal with the fact that the addiction to the use of substances is a consequence of a choice to use drugs or alcohol as a means to cope with unpleasant feelings due to a variety of causes. The use itself has a function. When an addict chooses to have treatment in a drug-free therapeutic community, this means a choice to stop using drugs and alcohol and to face the unpleasant feelings that are no longer avoidable by the use. This usually results in a dilemma of choosing between the unpleasant consequences of continuing to use and the unpleasant consequences of stopping. To make this choice is the first requirement for a successful treatment geared at a life independent of the use of drugs and alcohol. This choice is never an absolute one. There will always be an ambivalence.

What influences the choice to give up using or what influences the motivation to choose treatment? Negative results of therapy are often attributed to a lack of motivation in the client. However, several authors have stressed the importance of the therapist in the treatment of clients that are often labelled untreatable. It is crucial to transfer a message of hope to persons who have a 'failure identity' and have great problems in believing that treatment can be successful for them (Bratter, 1978; Walburg, 1984; Schaap, 1985). An important influence on the choice to seek treatment is the presence of pressure in the current situation. This can be pressure from partners, the family, the work situation or the threat of being arrested. It can also be the lack of the availability of drugs, physical health problems, housing problems, in short unpleasant conditions that make people choose another unpleasant condition, namely treatment.

A positive motivation to choose treatment not as a way to avoid an unpleasant situation but to choose a positive drug or alcohol-free life style.

This positive motivation is rarely present at admission to the therapeutic community but usually develops during treatment often several months after the onset of the treatment (Kooyman, 1975c).

Motivation mainly based on external influences can be called negative motivation. The use of drugs or alcohol is causing too many problems. A negative motivation to choose treatment can be that not choosing treatment could mean continuation of a prison sentence or other detention. In many countries, addicts can apply for treatment after their arrest. When the addict is not charged with serious criminal behavior, such as murder, drug dealing, assault using violence, he may be referred to an induction program of a therapeutic community. This possibility to choose treatment instead of prison however is not the same as compulsory treatment. The client has to make a choice. He can run away from the premises of the therapeutic community. The staff will not run after him, but if he does not return within 48 hours, the staff will inform the police.

Compulsory treatment

The wish to stop using drugs or to receive therapy does not usually come freely to an addict. There is usually some sort of external pressure. Legal pressure may be successfully utilized to bring some addicts to treatment (Brown et al., 1987). But it is questionable to what extent forced treatment is successful. Some authors are of the opinion that compulsory treatment is in any case better than no treatment at all. Ausubel (1972) for example sees addiction as an infectious disease. In dealing with epidemics of infectious diseases forced treatment is also used. An addict who is admitted to an institute would no longer function as a source of further infections.

Others say that addicts can not be changed by force. Heckman (1986) realizes that outside pressure, including legal pressure, may lead an addict to choose treatment. He states, however, that the application of legal pressure is often too strong and has an opposite effect. Heckman argues, that there is hardly any situation which kills motivation more than imprisonment. The important thing in prison is to survive within the restrictive structure. In addicts who are anxious to receive therapy during remand, imprisonment leads to rebellion and defensive reactions. In this way motivation can be destroyed in prison.

Compulsory treatment of drug addicts is not new. Before the Narcotic Addict Rehabilitation Act came into effect in 1962 in the U.S.A., forced treatment was almost the only treatment available. It was concentrated in two federal hospitals; one in Lexington and one in Fort Worth (Platt, 1986). The early results of follow-up research among ex-patients of the U.S. Public Health Hospital in Lexington showed some positive effects. Pescor (1943) found that 24% of the male patients who were discharged between January

1st 1936 and December 31th 1940, were drug-free at the time of the follow-up study. The best results were seen with those who had received compulsory after-care of any form (of the drug-free groups 55.7% were parolees and 38.6% probationers). In this research information was obtained about 60% of the total group of 4766 ex-patients. In the follow-up study of Hunt and Odoroff (1962) of 1912 male and female ex-patients from New York of the Lexington Hospital who were discharged between July 15th 1952 and December 15th 1955, information was obtained about 98.4% of the population. They found high relapse rates. However, less high rates were seen for those who had received compulsory after-care (91.2% vs. 85.7%). In a twelve year follow-up of 100 male New York addicts who had been hospitalized in Lexington between August 1952 and January 1953 Vaillant (1966) found a relapse to drug abuse in 90% of the cases during the period of the research. However, 46% had been found clean at the follow-up contact or at the moment of their death. Of those who had been treated on a voluntary basis, 96% had relapsed within a year after their discharge. Of those who had been admitted for at least nine months with one year compulsory after-care after their discharge, 67% had stayed drug-free for at least one year after their discharge.

The above described results do not provide an answer to whether the successes were due to the treatment or not. In 1973 the Lexington and Fort Worth hospitals were closed. The closure of these hospitals was not only due to the relatively poor results of the treatment. At the end of the Sixties, community-based treatment programs developed. There was a strong attitude in favor of de-institutionalization. A growing number of drug-free therapeutic communities had been founded where addicts could be admitted voluntarily or referred to by the justice system. Also large scale out-patient methadone maintenance programs had been set up as an answer to the increasing drug problems.

As indicated by the results of follow-up studies of ex-patients of Lexington, compulsory aftercare seems to have made a difference compared to no compulsory aftercare. In an article reviewing the results of treatment programs within federal prisons made possible through the Narcotic Addiction Rehabilitation Act, Petersen (1974) stated that variables related to the treatment had hardly any correlation with success after discharge. However, it was found that for some addicts, compulsory after-care consisting of counselling appointments with probation officers, had been more useful than a completely voluntary approach after discharge. Petersen concluded that compulsory inpatient treatment combined with compulsory outpatient treatment seems to be very effective in rehabilitating the drug addict (Petersen, 1974). In his description of the civil commitment program in California, Kramer (1972) attributes the poor results of his ambulatory program among other things to the fact that it is part of the justice system and not of the health system and because of this the program was not flexible enough to apply new treatment methods. In the last two decades more and more possibilities were created in the U.S.A. to refer addicts to treatment centers before they were sentenced. This is especially applied to minors and first-time offenders who can be referred for treatment by the District Attorney (Brown et al., 1987).

From their start, the drug-free therapeutic communities in the U.S.A. have treated drug-addicts referred for treatment by the justice system. The addict has a choice to choose between treatment and/ or incarceration. Although restricted the client still has some freedom of choice (Deitch and Zweben, 1979). He can choose to leave the program at anytime after which the police will be notified. Although there is definitively strong legal pressure present it is different from compulsory treatment in which the addicts do not have a choice but are just told to follow a treatment program. The treatment program is located outside the justice system's premises. The treatment program in the therapeutic community is followed together with 'voluntarily' admitted residents. The treatment results do not seem to differ in these two categories of residents (Rosenthal, 1977; De Leon, 1987).

The drug-free therapeutic communities also developed therapeutic communities within prisons. Addicts sentenced to prison detention could apply to be admitted to these programs (Chinlund, 1978; Wexler, 1986). Most of these programs were closed after some time. They had faced problems such as being insufficiently separated from the prison climate, being misunderstood by prison personnel working in other parts of the prison, having inadequate re-entry facilities and insufficient aftercare. In 1977 a system was set up in New York partly within the prison setting (during the initial phase), partly outside the prison setting in the existing therapeutic communities of seven large treatment organizations (the following phases). This program called 'Staying Out' has created an opportunity to proceed gradually from the prison system to the treatment system of a therapeutic community outside the prison.

The therapeutic communities in the Netherlands nowadays do receive residents referred by the justice system. The addicts have to apply for this themselves. They can be admitted from the prison to the therapeutic community before they are sentenced. Their court case is suspended for a period of four to five months after which the case will be dropped or a suspended sentence will be given if the resident is still in treatment. Another possibility is that after conviction the last half year of the prison sentence can be spent in a therapeutic community by a special arrangement. Addicts in the Netherlands can be committed to treatment in a psychiatric hospital when they are seen as dangerous to themselves or others and regarded as not responsible for their behavior due to their mental conditions. However, this law, meant for psychiatric patients, has hardly been used in recent years.

In Germany, special institutes were created for compulsory treatment. Compulsory programs must offer an alternative to drug abuse otherwise they will have no effect. They can be seen as settings with intensive care. When residents stay long enough to benefit (at least nine months until regularly discharged) over half of them seem to have improved considerably (Paschelke et al., 1986). Jongsma (1986) stated that in his opinion, centers of this kind where treatment is completely compulsory can only be effective if the compulsion can be combined with the safe limiting and constraint of concerned parent figures. Setting limits without this context of stimulating love will inhibit instead of stimulate growth and well-being.

Scholer (1986) described how the therapeutic community Persat Pertolongan in Malaysia changed from fighting for the rights of the drug dependent to seek his voluntary recovery into accepting compulsory treatment. Because most addicts were brought in still intoxicated and unable to take any decision the staff changed their policy completely. Compulsory admission brought unmotivated addicts to treatment and kept them in long enough to be motivated by the staff and fellow-residents. The legal system relieved the staff having to spend time keeping the residents in treatment or convincing parents that it was not yet the right moment to take their children away from the program. However, it is essential that the staff be aware of the necessity to actively motivate the residents to choose a positive lifestyle and not to regard it as sufficient that the residents only stay in the program. Scholer compares involuntary treatment with parents sending young children to school, this is also compulsion, but compulsion with concern.

Compulsory programs may be an answer for some addicts who are very destructive to themselves and others. Working in compulsory programs puts great demands on the staff. Compulsory treatment will, however, never be a solution to the whole addiction problem. Even in controlled situations with enforced treatment for all drug-addicts, some hard core addicts keep relapsing or they change to addictive behavior that is legally accepted. This is the case in Singapore where a system has been developed in which all drug addicts that are caught by the police have to spend at least 6 months in an enforced treatment institution.

On the other hand, compulsory treatment may have better results than no treatment at all. As long as waiting lists exist for voluntary treatment such as the therapeutic communities for addicts in Amsterdam, compulsory treatment does not seem to be a priority. Apart from the possible advantage of compulsory treatment, to get individuals into therapy who otherwise would have continued their drug use, there are serious disadvantages especially of compulsory treatment in a therapeutic community system. A therapeutic community is like a pressure-cooker. It produces stress in the individual who can no longer respond by using drugs or alcohol, acting-out behavior or isolating himself. The only possible response dealing positively with the pressure in a therapeutic community, namely running away, is no longer available in compulsory treatment. Temporarily leaving a therapeutic community by the resident may be necessary as a relief from pressure that seems unbearable. When this is impossible, the situation is like an engine without a safety valve. This means that in a closed system of involuntary treatment in a therapeutic community from which the resident cannot run away, watching the pressure in the system carefully becomes necessary to ensure that it is not too high.

When a resident is referred from the justice system to a normal therapeutic community after having made a choice between this option and staying in prison, this disadvantage does not exist. The resident can always decide to discontinue the treatment although this may mean a possible arrest with the consequence of having to spend time in prison. There is here a great difference compared with enforced treatment ordered by court decisions leaving out the responsibility of the addict to make a choice. Taking the choice of treatment instead of prison is not so different from taking the choice of treatment instead of being fired from a job, separated from a partner or expelled from the home (Kooyman, 1977).

Improving the motivation

The treatment process after admission to a therapeutic community can improve the motivation to stay in treatment and change a negative motivation into a positive one. Also the motivation, usually a negative one to start with, can be influenced by the attitude of the therapist or counsellor before admission. The following three factors are important:

1. Pointing out the negative side; the misery of the addiction. This means confronting the addict with his denial, pointing out the reality of the situation, often a painful one.
2. Creating hope. The presence of recovered addicts in the introduction team in itself can underline the message that successful treatment is possible.
3. Making demands of the clients. By asking the clients to show their motivation by performing simple tasks, the message is given that they can do something themselves.

Also treatment tends to be more attractive when the addict has to do something to get admitted. An illustration of this is the experience of the Emiliehoeve. More clients who came for the orientating interviews for admission to the Emiliehoeve continued their induction when they were asked to do something to get admitted, than during the first year when friendly staff members tried to convince the addicts that it was important for them to come to the therapeutic community. By asking the client why he wanted to stop, the inductee may not have given honest answers, but this certainly made the client act as if he really wanted to stop using. By explaining why, the motivation to stop was reinforced. The most effective way to introduce clients to a therapeutic community by improving their motivation for treatment is using group induction meetings run by staff and assisted by older residents of the therapeutic community.

Jongsma and Van der Velde (1985) showed that the so-called high threshold of the therapeutic communities is a myth. They have found that the same type of addicts are in treatment in the therapeutic community as in the so-called low threshold program, which puts few or no demands on the clients to be admitted. 1985).

Indications and contra-indications of the treatment

For some addicts, therapeutic communities can be the treatment of choice. For others this treatment may be no solution or this treatment may even be contra-indicated. Cancrini (1985) suggests in his study of drug addicts and their families that addicts from multi-problem families, often children of economically and culturally deprived women, who have often been abandoned in public institutions, can best be treated in therapeutic communities. Addicts who have adapted to a junky life-style need an intensive 24-hour treatment setting to restructure their lives and acquire new personal values necessary to lead a life free from drugs and crime. Addicts that have still kept a positive social life, who still have jobs and a healthy family support system, may not need an extensive treatment program such as a therapeutic community. They may be well enough off in an ambulatory day or evening program. Involvement of the family or partners in these programs however, seems crucial for a successful outcome (Kooyman and Van Steijn, 1990).

The treatment in a therapeutic community is in principle the same for all residents. This in itself limits the therapeutic possibilities. The community itself is the main therapeutic element. Not all residents require the same treatment and not all addicts fit into one type of therapeutic community program. One model of a therapeutic community cannot be the answer for all addicts. There are different types necessary for different addicts. Little is known of the limitations of specific therapeutic community models. For some sub-groups of addicts, special therapeutic communities are created. There are special therapeutic communities for adolescents, offering more possibilities for education, sports and recreational games than regular therapeutic communities. Special therapeutic communities have been created for minority groups: blacks, hispanics and homosexuals. Special therapeutic communities offer treatment for addicts with specific psychiatric problems such as borderline personalities. There are therapeutic communities for women only. Some women have been brutalized and hurt by men in such a way that treatment amongst men seems undesirable, at least in the initial phase.

Contra-indications for treatment in a therapeutic community can be divided into absolute and relative ones (Schaap, 1987). Admission to a therapeutic community is absolutely contra-indicated for the following kind of addicts:

- patients with cerebral defects or lack of intelligence which makes them unable them to understandthe concepts of the program . A resident does not have to be bright or intelligent, however. Most addicts that survive in the street have sufficient intelligence to follow the program in a therapeutic community.
- patients with a psychosis, such as schizophrenia, paranoid states or a panic or manic depressive psychosis. There must be enough reality- testing possible and the ability to be held responsible for one's behavior.

Relative contra-indications for admission are:

- extreme sociopathic behavior with an inability to form object-relationships.
- persons that have relapsed after completing the program. They may need diffe-rent treatment. For such persons it is often very hard to overcome the feelings of having failed when they have relapsed after completion of the program. They may do better in another therapeutic community or another treatment modality with different staff.
- serious physical impairment making participation in certain activities of the community impossible.
- lack of alternative treatment. More appropriate treatment such as day treatment or a short term program is not available.
- age: when the age group is mainly between 18 and 35, older persons and younger persons may not have sufficient peers to identify with.
- language: in most cases speaking a different language may only be a temporary problem. The therapeutic community is an environment in which languages can be learned and practised easily.

Separate therapeutic community programs adapted for the needs of specific addict populations such as clients with AIDS or borderline personalities have to be developed (Osterhues,1990). As each therapeutic community offers an approach similar for all residents this program can not serve the specific needs of all addicts. Apart from special therapeutic communities for special target groups such as women, adolescents, alcoholics, minority groups, few definite answers are given on questions such as what type of addicts are best helped by what type of therapeutic communities or for what type of addicts is a therapeutic community program the best type of treatment. Besides the above-mentioned psychiatric conditions, there are few contra-indications for admission to a therapeutic community. Suicidal or auto-mutilative tendencies are no contra-indication. In fact, this behavior seems to disappear after admission to the structured environment of the therapeutic community.

To benefit sufficiently from the treatment in a therapeutic community the client has to participate for a long period in the program. As this study demonstrates, the length of time spent in the program is the main determinant for successful outcome. Persons with a tendency to drop-out and who are not yet fully convinced that they have to stop using drugs for the rest of their life will have less favorable outcome results. Therapeutic communities are not the only answer to the drug problem. There are other ways for people to overcome their addictions. They are, however, a solution for many addicts, especially those who do not have any positive factors in their environment.

CHAPTER 9

Treatment outcome literature

The therapeutic communities for addicts and research

The therapeutic community is a relatively young member of the care scene. It emerged to fill a gap; a need that the professional institutes could not provide for those individuals who were alienated from 'normal' society. Being alienated is an important and often neglected element in the recovery process. In order to grow, like any maturing being, objective knowledge of itself is necessary. A little distance from its commitments is needed; it can not withdraw into itself without danger of extinction. In the early stages of the building of a therapeutic community, a lot of creativity is needed. However, to maintain a community after it has been established, requires something more than constant creativity. To maintain a therapeutic community requires accountability and the balancing of subjective with objective factors. This balancing has already been set in motion in the American therapeutic communities with the introduction of professionals into the earlier self-help process. A further step was the introduction of scientific research to explain what is going on, not only to the therapeutic community, but also to the outside world. The sympathetic appeal to people to believe in the therapeutic community concept is not enough. A clear understanding brought about by a rationally designed project is needed too. There have perhaps been too many signs of the religious zeal and of cult phenomena in the therapeutic community movement.

Many therapeutic community members distrust science. The therapeutic community works. Staff often cannot accept that the program does not work for all residents. By not accepting the possibility of failures and suggesting a 100% success rate, one sets it up to be judged by a logic of absolute ends rather than a logic of consequence and responsibility. Science rejects absolute thinking and instead states its case in probabilities and contingencies. For example consider the measurement of 'success rate'. Many therapeutic community members secretly fear that their success does not approach the absolute end of 100%. They fear that if they admit something considerably lower, they endanger their existence by admitting to failure. But the scientific method intervenes with the question: „What is the probability of success in the first place?" Addiction, like other chronic diseases with which it can properly be compared, has a relatively low success rate no matter what treatment method is used. If you are helping 30% of the population that entered your door to stay off drugs for at least 3 years after treatment, you are doing your job well. But many therapeutic community members do not know this, because they fear they are not up to 100%. The old justification that it is maybe helping 'just one addict', is the other side of the same religious coin. Scientific research shows what is really possible and therefore provides an objective standard for the community to account for itself, to itself and to a sceptical world.

The introduction of science can preserve the experience of the therapeutic community movement. 'Hard' data of its achievement as well as its failures can provide a science of itself and rational claims towards its universality. Without science, the therapeutic communities can easily become something nostalgic ('remember the good old days'), something that was bound historically to a particular place and time and therefore has outlived its usefulness being a simple fad or fashion. As the therapeutic community concept matures, a sense of history needs to develop. Science must work with data and documents. The therapeutic community, governed partially scientifically, must organize its documents and records in order to gain an overview of itself. This is not merely administration for the record, it should also be able to provide the answer to questions raised. Therapeutic Communities should not only go back into an exploration of their origins in a systematic way -through 'oral history' for instance- but should do long-term follow-up investigations of its members over an entire life span (Kooyman and Kaplan, 1986).

Outcome research in the United States

During the last twenty years, many studies have been done to measure the outcome results of therapeutic communities. Follow-up studies were carried out using face to face interviews held with former residents of therapeutic communities. Many of these studies only included individuals who had completed the full treatment program. Collier (1970), for instance, interviewed graduates of Daytop Village in the late Sixties. Of the 38 persons interviewed one year after they had completed the program, he found that only 4 had relapsed into drug abuse; 11 of the 38 cases however had been employed as staff in the treatment program of Daytop Village; 17 had other jobs, one had a job at another drug rehabilitation program and the occupation of 5 individuals was unknown. Casriel and Amen (1971) also did a follow-up study of Daytop graduates: 90% of them had not relapsed in the year after they left the treatment program, 25% had jobs in other fields than drug-rehabilitation. In 1971 a larger group of graduates of Daytop Village were involved in a follow-up study. Again it was found that 90% of the 250 graduates had not relapsed. In this study a group had also been interviewed which left prematurely but at least 6 months after admission to the program. The relapse rate was 50%; however 41% of the graduates and 63% of the persons that had left prematurely after 6 months could not be traced (Collier, 1971). Collier repeated the follow-up research in 1972: 84% of the graduates had not relapsed into drug use, had no alcohol problems, no arrests and were working or at school; 46% of the residents who had left prematurely after 6 months did fit into those criteria (Collier and Hijazi, 1974).

De Leon (1984) developed an extensive research program in Phoenix House, New York. He made charts of a five year period prior to admission in the therapeutic community and compared this period with a five year period after discharge. Cohorts of admissions were studied including not only individuals who had completed the program or who had stayed at least 6 months, but also early drop-outs (De Leon et al., 1982). In a two year follow-up De Leon found for all clients an improvement in items such as 'drug use', 'criminality' (arrests) and 'employment'. When success was defined as no use of heroin while being employed for at least 75% of the time after discharge and no criminal behavior, he found a significant relationship between success and the time spent in the program. In the two year follow-up of two different cohorts of Phoenix House residents, De Leon found a strikingly similar relationship of time spent in the program with success and improvement (De Leon and Jainchill, 1981).

Several therapeutic community programs developed their own research programs during the 1970s: in the United States: the Phoenix House program by De Leon and Jainchill; the Daytop Village program by Biase and programs by Barr in Eagleville, and Holland in Gateway House (De Leon, 1984; De Leon and Jainchill, 1986a; Biase, 1986; Barr, 1986; Holland, 1983b). These researchers were part of the treatment organization. This can lead to a possible bias as their programs have had to prove success and this could have affected the results. Researchers, employed by the treatment programs may show an over-advocacy. On the other hand, these program-based researchers have the advantage that it is much more easier for them to get the necessary cooperation from residents and staff to implement their research.

In an overview of 19 studies on the effectiveness of the therapeutic community in the treatment of addicts completed between 1972 and 1982, Holland (1983a) summarized the degree of change reported in three areas: drug abuse, criminality and employment. Of these 19 studies, 13 were program-based (the therapeutic community programs of Abraxas, Daytop Connecticut, Daytop Village New York, Eagleville, Gateway House, Phoenix House and Teen Challenge)(see table 4.a). The remaining 6 were multi-modality studies comparing results of one or more therapeutic communities with one or more other treatment modalities, such as outpatient drug-free or methadone maintenance programs (see table 4.b).

In table 4.a the first column indicates the authors and date of the study. The second column shows the attempted sample of clients to be contacted or interviewed and the achieved sample. The smaller the difference between the number of clients, the more meaningful the evaluation results. The third column shows the contrast groups (early drop-outs versus graduates, short term versus long term). Only one of the single modality studies and one of the multi-modality studies did not break the total sample into contrast groups. Contrast groups help to test the treatment hypothesis in the absence of a treatment control group. The fourth column lists the measures used. Most have measures for the same behavioral criteria: drug and alcohol use, criminality, education, employment and psychological functioning. The fifth column shows the pre-treatment period of time being compared with the post treatment period of time. To avoid artifacts (if the period before and after treatment is too short the study will inevitably show improvements) minimal evaluations should show six months pre- and post treatment periods. Ideally the evaluations should be both short term and long term (i.e. up to five years) outcomes. The last column, 'additional analysis,' indicates if any non-treatment factors such as motivation, being under more pressure to change, explaining successful outcome were studied. Some studies are prospective in the way that they have withheld pre-treatment information at intakes and post-treatment information at follow-up. Others are retro-spective, both pre- and post-treatment information was obtained at follow-up. Some studies have post-treatment tests only.

Legend table 4a and 4b (Holland, 1983a)

DO	Left before completing treatment.	GRAD	Completed treatment.
LT	Long-term.	ST	Short-term.
TIP	Time in program.	TOP	Time out of program.

Note. A dash or NA means not provided or not applicable.

a Interviewed subjects included alcoholics (N=162) and drug addicts (N=111); outcomes are presented here only for drug addicts.
b Interviewed subjects included alcoholics (N=99) and drug addicts (N=93); outcomes are presented here only for drug addicts.
c Outcomes for a second (validation) cohort are described in the report but are not presented here.

Table 4.a: Posttreatment evaluations of therapeutic communities 1972-1982; single modality studies (Holland, 1983).

Study	N Achieved (Target)	Contrast Groups	Measures	Baseline and Follow-up Periods	Outcomes	Additional Analyses
DeLeon, Holland, & Rosenthal (1972)	358 --	5 DO groups (<3 mo., 3-5 mo., 6-8 mo., 9-11 mo., 12+ mo.) & "remained" (TIP [remained]=22 mo.).	Agency data: single-item measure of criminality	1 yr. preadmission, during treatment, & 1 yr. postdischarge	For DO groups, 1, 2-4, & 5, respectively, percent arrested decreased -7%, -40 to -50%, -70%, & -90%.	--
Barr, Rosen, Antes, & Ottenberg (1973)	273 (724)[a]	3 treatment phase groups: DO residential (TIP=14 da.); GRAD residential (TIP=3 mo.); some REENTRY (TIP=8 mo.).	Self-report & other: single-item measures of drug & alcohol use, source of food & shelter.	Posttest only: lifetime postadmission (drugs & alcohol), & at time of follow-up (source of food & shelter)(11 to 34 mo. postadmission, TOP =24.7 mo.).	For treatment phase groups DO, GRAD, & REENTRY, respectively, abstinent 6+ mo. since discharge = 22%, 31%, 62%; productively employed at follow-up = 32%, 45%, 67%; abstinent 6+ mo. & productively employed = 18%, 29%, 57%.	Some outcomes compared with status at intake.
Collier & Hijazi (1974)	272 (552)	DO (minimum 6 mo. in treatment, TIP=12 mo.), & GRAD (TIP= 22 mo.).	Self-report & other, mixed retro- & prospective: single-item measures of drug use, criminality, employment, treatment for abuse, social adjustment, treatment satisfaction.	Lifetime preadmission (arrests, income); lifetime postdischarge (drugs, arrests, income, treatment); & at time of follow-up (employment)(6 to 96 mo. postdischarge, TOP=12 mo.).	46% of DO & 85% of GRAD not using drugs, not arrested, & employed/in school at follow-up. Outcomes for DO who remained 12+ mo. more favorable than outcomes for DO who remained 6-12 mo.	--

Table 4.a. continued

Study	N Achieved (Target)	Contrast Groups	Measures	Baseline and Follow-up Periods	Outcomes	Additional Analyses
Romond, Forrest, & Kleber (1975)	40 (--)	DO (TIP=6 mo.) & GRAD (TIP=21 mo.).	Self-report: single-item measures of drug use, criminality, employment, treatment for abuse.	Posttest only: lifetime postadmission (6 to 47 mo., TOP [DO]=27 mo., TOP [GRAD]=17 mo.).	For DO & GRAD respectively, % of TOP employed/in school = 40%, 94%; % of TOP addicted = 35%, 0.5%; % of TOP incarcerated = 29%, 0%. Among DO, increases in TIP associated with more positive outcome.	Follow-up period adjusted for time at risk.
Pin, Martin, & Walsh (1976)	200 (300)	2 TIP groups: ST (<3 mo.) & LT (12+ mo.) (TIP NA).	Self-report: single-item measures of drug use, criminality, employment.	Posttest only: at time of follow-up (TOP NA).	For ST & LT respectively, using no "hard" drugs = 26%, 96%; not arrested = 43%, 77%; employed = 45%, 93%.	Client variables regressed on TIP, type of discharge, outcome.
National institute on Drug Abuse (1977)	186 (366)	3 treatment phase groups: DO induction (TIP=14 da.); DO residential (TIP=3 mo.); GRAD residential (TIP=7.6 mo.).	Self-report, retrospective: single-item measures of substance use, criminality, employment, treatment for abuse, religiosity.	At time of admission & at time of follow-up (7 yr. postadmission).	For DO induction, DO residential, & GRAD respectively, % using heroin decreased -79%, -98%, -95%; % arrested decreased -1%, -24%, -64%; % received treatment for abuse increased +100%, +65%, & decreased -52%.	--

Table 4.a. continued

Study	N Achieved (Target)	Contrast Groups	Measures	Baseline and Follow-up Periods	Outcomes	Additional Analyses
Holland (1978a)	193 --	2 DO groups (<9 mo., TIP=3 mo.; 9+ mo., TIP=17 mo.) & GRAD (TIP=28 mo.).	Agency data: Single-item measures of criminality.	1 yr. preadmission & 1 yr. postdischarge.	For DO groups 1 & 2, & GRAD, respectively, arrest rate decreased 0%, -81%, -97%.	Outcome examined by type of arrest, sex, race, age.
Holland (1978b)	400 (684)	3 DO groups (<3 mo., TIP=1 mo.; 3 to <9 mo., TIP=5 mo.; 9+ mo., TIP=18 mo.) & GRAD (TIP=26 mo.).	Self-report, retrospective: single-item measures of drug use, criminality, employment, treatment for abuse.	2 yr. preadmission & 2 yr. postdischarge.	For DO groups & GRAD, respectively, % using opiates decreased -72%, -75%, -90%, -96%; no. of arrests decreased -34%, -50%, -74%, -98%; no. employed increased +16%, +117%, +106%, +160%.	--
DeLeon, Andrews, Wexler, Jaffee, & Rosenthal (1979)	202 --	6 TIP groups: <1 mo., 4-6 mo., 8-10 mo., 12-14 mo., 15-19 mo., 20-26 mo.	Agency data: single-item measure of criminality.	3 yr. preadmission, 3 yr. postdischarge, & lifetime postdischarge (3 to 6 yr.).	Pre-post decreases in percent arrested & arrest rate occurred as a function of TIP.	Follow-up period adjusted for time at risk. Outcomes examined by legal status, race, sex.
Pompi, Shreiner, & McKey (1979)	223 (250)	None (TIP[all]=12 mo.)	Self-report, prospective: single-item measures of drug & alcohol use, criminality, employment, education, treatment for abuse, suicide; & composite measures of productivity, drug involvement.	1 yr. preadmission & 1 yr. postdischarge.	Pre-post decreases in no. of arrests (-73%); no. in jail (-64%); % using 8 out of 10 substance (-50% to -95%); increase in no. employed (+105%).	--

Table 4.a. continued

Study	N Achieved (Target)	Contrast Groups	Measures	Baseline and Follow-up Periods	Outcomes	Additional Analyses
Barr & Antes (1982)	192 (292)[b]	3 treatment phase groups: DO residential (TIP=24 da.); GRAD residential (TIP=61 da.); some REENTRY (TIP=197 da.).	Self-report: single-item measures of drug & alcohol use, criminality, employ-ment, education, medical, psychologi-cal; & composite measure of success.	Posttest only: lifetime postdischarge (79 to 87 mo., TOP= 84 mo.).	During follow-up period, DO averaged 20 mo. in good status, GRAD aver-aged 32 mo. in good status, & REENTRY averaged 62 mo. in good status.	Follow-up period adjusted for time at risk, Pre-, during-, & posttreatment vari-ables regressed on outcome status.
DeLeon, Wexler, & Jainchill (1982)	273 (307)[c]	6 DO groups (<1 mo., 1-4 mo., 5-8 mo., 9-12 mo., 13-16 mo., 17+ mo.) & GRAD (TIP NA).	Self-report & other, retrospective: com-posite measures of drug use, criminality, employment, success.	1 yr. preadmission; 1, 2, 3, 4, & 5 yr. postdischarge; life-time postdischarge (TOP[DO]=4.7 yr., TOP[GRAD]=6.4 yr.).	For DO & GRAD, respectively, pre-post increase in "success" = 31%, 75%; pre-post increase in "improve-ment" = 56%, 93%. Among DO, increases in TIP associated with more positive out-come.	Validation cohort. Outcomes examined by sex, race, primary drug.
Holland (in press)	400 (648)	3 DO groups (<3 mo., TIP=1 mo.; 3 to <9 mo., TIP=5 mo.; 9+ mo., TIP=18 mo.) & GRAD (TIP=26 mo.).	Self-report & other, retrospective: com-posite measures of drug use, alcohol use, criminality, employ-ment, psychological, social stability.	Lifetime & 30 da. preadmission; lifetime postdischarge; 30 da. pre-follow-up (2 to 10 yr. postdischarge, TOP=56 mo.).	Pre-post improvement for all groups on all measures except alcohol. Degree of improvement a func-tion of TIP for drug use, criminality, em-ployment, & psycho-logical.	Other explanations for TIP effect examined, including selection, maturation, differential completion rates, differential TOP, & differential validity of self-report data.

Table 4.b: *Posttreatment evaluations of therapeutic communities 1972-1982; multi modality studies (Holland, 1983).*

Study	Contrast Groups	N Achieved (Target)	Measures	Baseline and Follow-up Periods	Outcomes	Additional Analyses
Nash (1973)	Residential Drug Free	198 (308)	Agency data & other: single-item measure of criminality	Lifetime preadmission & lifetime post-admission (mean time post-admission=17 mo.).	For MM & RDF, respectively, percent arrested decreased -60%, & -60%; arrest rate decreased -23% & -33%. TIP associated with more positive outcomes for RDF but not for MM.	Outcomes analyzed by type of charge, client characteristics, & program characteristics.
	Methadone Maintenance	249 (269)				
Burt & Associates (1977)	New York City Programms:[a]		Self-report, mixed retro- & prospective: single-item measures of drug use, criminality, employment; composite measure of success.	60 da. preadmission, 60 da. postdischarge, & 60 da. pre-follow-up (3 to 4 yr. postadmission).	Significant pre-post improvement by all groups, including no treatment. At follow-up, 50% of all clients fully recovered, 33% partially recovered, 10% marginally recovered, & 7% failure. TIP not associated with outcome.	Pretreatment client variables regressed on outcome.
	Residential Drug Free	185 (290)				
	Outpatient Drug Free	135 (219)				
	Methadone Maintenance	142 (273)				
	No Treatment	118 (223)				

Legend:

LT	Long-term.	RDF	Residential Drug Free.
MM	Methadone Maintenance.	ST	Short-term.
OD	Outpatient Detox.	TIP	Time in Program.
ODF	Outpatient Drug Free.	VA	Veterans Administration.

Table 4.b. continued

Study	N Achieved (Target)	Contrast Groups	Measures	Baseline and Follow-up Periods	Outcomes	Additional Analyses
Simpson, Savage, Lloyd, & Sells (1978)	Residential Drug Free Outpatient Drug Free Methadone Maintenance Outpatient Detox No Treatment	856 (1,097) 425 (567) 1,485 (1,904) 236 (323) 162 (216)	Self-report, mixed retro- & prospective: single-item measures of drug use, criminality, employment, treatment for abuse, months unsupervised; composite measure of success.	Lifetime preadmission (arrests, jail, treatment); 1 yr. preadmission (employment); 60 da. preadmission (drug use, alcohol use, employment); 1 yr. postdischarge.	RDF, ODF, & MM had more favorable outcomes than OD & no treatment with respect to drug use & employment. MM had the most favorable outcome with respect to jail; & RDF & ODF has most favorable outcomes with respect to treatment for abuse. TIP associated with outcome for RDF (8 out of 10 criteria); for ODF (5 out of 10); & MM (4 out of 10).	Outcomes adjusted for time at risk, and for client characteristics (ANCOVA). Client characteristics regressed on outcome.
Veterans Administration (1979)	None[b]	1,471 (2,269)	Self-report, prospective: single-item measures of drug & alcohol use, criminality, employment, medical, social, psychological, treatment satisfaction.	30 da. preadmission & 30 da. pre-follow-up (44 mo. postadmission).	Clients showed significant pre-post improvement with respect to all outcomes. TIP unrelated to outcome except medical, & this relationship was negative.	Client variables and TIP regressed on outcome.

[a] This study also reported outcomes for a sample of Washington, D.C., programs, which did not include residential drug free (Therapeutic community).

[b] Outcomes by program modality not reported.

[c] Two-thirds of Methadone Maintenance clients were in treatment at time of follow-up.

[d] Outcomes by program modality not reported. Interviewed subjects included drug addicts (N=282; ST=57, LT=225) and alcoholics (N=460; ST=98, LT=362). Only outcomes for drug addicts are presented here.

Table 4.b. continued

Study	N Achieved (Target)	Contrast Groups	Measures	Baseline and Follow-up Periods	Outcomes	Additional Analyses
Holland (1978a)	193	2				
Bale, Van Stone, Kuldau, Engelsing, Elashoff, & Zarcone (1980)	Short-term Residential Drug Free (TIP=21 da.): 75 — Long-term Residential Drug Free (TIP=1 yr.): 75 — Methadone Maintenance[c]: 59 — Outpatient Detox: 224 — Non-VA Treatment: 112 — Total: 545 (585)		Self-report, prospective: single-item measures of drug use, criminality, employment; composite measure of success.	Posttest only: 1 yr. postadmission (arrests, convictions, employed); 30 da. pre-follow-up (1 yr. postadmission).	LT RDF & MM had more favorable outcomes than non-VA treatment with respect to heroin use, convictions, time in jail, employment. LT RDF had more favorable outcomes than non-VA treatment with respect to use of other drugs, arrests. TIP associated with more positive outcomes.	Outcomes adjusted for client characteristics (ANOVA & ANCOVA).
McLellan, Luborsky[d], O'Brien, Woody, & Druley (1982)	Short-term (6 to 15 da.): 155 — Long-term (15+ da.): 587 — Total: 742 (879)		Self-report, prospective: single-item & composite measures of drug use, alcohol use, employment, criminality, family, medical, psychological.	30 da. preadmission & 30 da. pre-follow-up (6 mo. postadmission).	ST showed significant pre-post improvement with respect to drugs, employment, & psychological. LT showed significant pre-post improvement on all measures but medical. LT had more favorable outcomes than ST on all measures.	Outcomes adjusted for client characteristics & pretreatment criterion scores (ANCOVA).

A few evaluation studies have utilized a composite index of successful outcome, combining measures of criminal activity, drug use and employment. With respect to such global measures of success, outcome ranges from 20% totally successful among early drop-outs to 85% totally successful among graduates (Holland, 1983a; De Leon, 1982, 1984). In the area of drug abuse measured in terms of the reduction in percentage of former residents using drugs at follow-up or the percentage of time since discharge having used drugs, these studies show 25 to 40% of the early drop-outs (i.e., less than 90 days treatment in the program) not using drugs, compared to 85 up to 90% of the graduates not using drugs. In the area of criminality, measured in terms of the number of arrests or convictions, these studies show post-treatment reductions in criminality, ranging from no change for early drop-outs to 98% reduction in arrests among graduates. In the area of employment, measured in terms of the increases in percentage of former residents employed or percentage of time employed, there is a pre-post treatment increase in employment from 30 to 45% among early drop-outs to an increase of time employed by over 150% among graduates (Holland, 1983a).

Time in program predictor of success

The most consistent predictor of successful outcome has been the length of stay in treatment. Research also indicates a gradual difference in improvement instead of an absolute difference between completions and drop-outs (Holland, 1983a; De Leon, 1984; Berglund et al., 1991). Graduates are significantly better than drop-outs on all measures of outcome (De Leon, 1989). Among drop-outs there is a positive relationship between outcome and length of stay in treatment (Barr and Antes, 1980; De Leon, 1984, 1985; Holland, 1983; Coombs, 1981; Wilson and Mandelbroke, 1978). Comparing the retention and improvement rates of two cohorts of residents admitted at Phoenix House, New York, De Leon found virtually indistinguishable results comparing the cohort from 1970-1971 and the cohort from the 1974 admissions (De Leon, 1985). There was 93% success among program graduates compared to 89% success among the longest staying drop-outs. Length of stay and successful treatment, however, may not be correlated for all residents (De Leon, 1989).

McLellan found a positive correlation between treatment duration and improvement related to treatment outcome in both a methadone maintenance program and a therapeutic community for clients with low or moderate levels of psychiatric problem severity, but a negative relationship in the therapeutic community for clients with severe psychiatric problems. These types of clients did not change in the methadone program (McLellan et al., 1984). This study has, however, been criticized as statistically invalid. There had only been 28 high severity clients in the therapeutic community sample (which in fact was not a therapeutic community in the concept tradition, but an in-patient program with a duration of only 60 days) (Holland, 1987). Holland (1987) found that psychologically sicker clients do not do as well in treatment in the therapeutic community as psychologically healthier ones. This finding is consistent in prior research in all forms of treatment. On the other hand these sicker residents did not get worse during treatment.

Coombs compared outcome results of a short-term therapeutic community (3 months) with those of a long-term therapeutic community (11-18 months). The most impressive behavior changes occurred among those who participated in the long-term program as compared to the shorter and among those who graduated as compared to drop-outs (Coombs, 1981).

Relation of client characteristics with successful outcome

Attempts to determine client characteristics indicative of subsequent treatment success, have not been very revealing. A review of program-based research found certain client characteristics have consistently correlated with the length of treatment, although the predicted power of these characteristics has not been corroborated in replicated study designs (Rosenthal, 1984). In the large scale American DARP studies in which 25 agencies participated, no evidence was found for improving client outcomes through an optimal match between client types and treatment types. Although some client characteristics, particularly lower pre-treatment criminal involvement, were related to more favorable outcomes, this was the case within each treatment type (Simpson et al., 1978; Joe et al., 1983). Age, race and other demographic factors have not been found to relate to outcome in therapeutic communities in most studies. In Phoenix House, females however, showed higher success rates and females revealed significantly better psychological results (De Leon and Jainchill, 1981).

Several correlates of positive outcomes of drug use, criminality or employment have been identified. Those include: lower life-time criminality, higher pre-treatment educational level (lower drop-out rates at school), opiates as a primary drug, less re-entry into treatment within the first postdischarge year (De Leon, 1989; Simpson and Sells, 1981). These correlations, though significant, were small compared to the effects of time in the program. They do not consistently predict successful states, measured by the compositive index (De Leon, 1985).

Persons who completed the Phoenix House Therapeutic Community program had been on average 25.3 years old at admission, compared to dropouts who had been 21.7 years upon entering the program. Persons who were younger at admission had been less successful. Black residents were doing slightly better than white residents. Contrary to Daytop Village, black residents have always been in a majority in Phoenix House, New York. Minority groups such as Hispanic residents (mostly of Puerto Rican origin) had less successful outcome results. Persons that had been referred by the courts did better than persons who were admitted voluntarily. There had been fewer early splitters among them (De Leon et al., 1982). Higher levels of psychopathology scores on psychological tests, such as the MMPI were also found to be related to less successful outcome (De Leon et al., 1973).

Sansone (1980) in a study of clients of Odyssey House, New York found lower retention rates for females, adolescents and Hispanic residents. The retention rate during the first six months in treatment of re-admission compared with single admissions was considerably higher especially for black male and older residents. Cutter (1977) found that previous education level at admission was related to successful outcome after treatment in a drug-free therapeutic community program of a state mental hospital as well as time spent in the program. In a NIDA study on predicting retention and follow-up status in the therapeutic community, De Leon (1984) found as significant predictors of success: lifetime criminality and psychological factors (less defensiveness, less denial, lower sense of responsibility). With the effects of the other variables removed, time in the program remained a significant predictor of success.

Retention studies

As the most consistent predictor of successful outcome is length of stay in the therapeutic community program, it is important to understand retention as a phenomenon. In the studies of retention, no clear client profile has emerged predicting length of stay in treatment. Some studies indicate that early drop-outs reveal higher psychological dysfunction as measured by standard psychological tests (Sacks and Levy, 1979; Wexler and De Leon, 1977; Zuckerman and Sola, 1975). De Leon (1973) found lower psychopathology scores at the MMPI tests of those residents staying longer than 6 months compared to those staying less than 6 months. Fourman and Parks (1981) found that MMPI scores significantly differentiate groups staying less than or more than 20 days in treatment, but had no power to predict more precise ranges of retention.

In a study assessing all therapeutic communities existing in New York in 1970, Winick (1980) computed the retention rates. He found a drop in retention rates, overall, of almost 14% from one to three months and a drop of 13% from three to six months. By three months almost half of the clients had left. After six months the drop decreases to 8% for the next three months, by twelve months the drop is 5%. Short-cycle programs had better retention than the long term programs. Therapeutic communities with a capacity not over 100 residents did better than larger ones. Sansone (1980) found in a study in Odyssey House in New York higher drop-out rates among female, adolescent and hispanic residents. She also found differences between re-admissions and first admissions. Males, blacks and older residents tended to have higher retention rates upon re-admission to the program. After six months these differences had disappeared. She suggests further separating first and re-admissions.

One year retention rates vary between 8% and 25% in most therapeutic communities, depending on the program studied. However, there are exceptions. The therapeutic communities of the Centro Italiano di Solidarieta in Rome with one long term and one short term therapeutic community claim a 90% retention rate. This therapeutic community has a long (up to 11 months) induction phase with ambulatory groups and extensive involvement of parents and significant others. Over two thirds of the persons coming for orientation (accoglienza) are not admitted to any of the therapeutic communities. Yohay and Winick (1986) reported a high retention capacity resulting in 85% of those who enter the program completing it in the AREBA program. This is a therapeutic community founded by Casriel applying the New Identity Process (bonding groups) in the treatment, based on the Synanon/Daytop model (Casriel, 1972).

Drop-out in all studies is highest within the first 15 days and declines thereafter. The majority of the drop-outs leave the therapeutic community within three months. Beyond 90 days the likelihood of continued stay in treatment increases significantly with longer stay in treatment (De Leon and Schwartz, 1984; Hendriks, 1990). The retention rates of most therapeutic communities show a characteristic pattern. Most drop-outs occur in the first four weeks (De Leon and Schwartz, 1984). Bschor (1986) compares the retention curves of the American therapeutic communities published by De Leon in 1984 with a retention curve of a German therapeutic community. As can be seen in Figure IX the hyperbola-like curves look very similar.

Drop-out reflects an interaction between client and treatment factors. Rather than static characteristics (such as demography or social background) the important client factors, such as circumstances (extrinsic pressures), motivation (intrinsic pressures) readiness and suitability for treatment, are dynamic (De Leon and Jainchill, 1986). Clients' differences in these factors however, emerge after their initial experience in the therapeutic community during the early days of residency (De Leon, 1989). In an attempt to improve the retention rate, an experimental study was carried out in Phoenix House, New York introducing three interventions:

1. Four 90 minute seminars a week, for two months for new admissions by senior staff members, addressing the therapeutic community philosophy and expectations and problems of staying in treatment.
2. Two 2 hour orientation sessions in the first two weeks with groups of the family members of new admissions and one session with a staff counsellor at the treatment facility.
3. Extra individual counselling sessions for new admissions during the

first 14 days of treatment.

There was a positive effect of the interventions administered separately and combined, particularly of the sessions for significant others (family-members) and the seminars by senior staff (De Leon and Jainchill, 1986b).

Family factors and retention

Drop-outs were more likely to come from a less deviant family (Wexler and De Leon, 1977). Aron and Daily (1976) reported more alcohol and drug abuse in the family of drop-outs. Condelli (1986, 1989) found more drop-outs among clients who had experienced less pressure for admission from significant others, before and during treatment.

In the above mentioned research in Phoenix House, it was found that attendance of seminars for parents during the early stages of treatment, reduced the drop-out significantly (De Leon and Deitch, 1985; De Leon and Jainchill, 1986b; De Leon, 1985). The extensive involvement of families in the therapeutic community program in Rome, may well contribute to the extremely high retention rate in the therapeutic community.

Legal pressure and retention

American therapeutic communities have served clients referred from the criminal justice system from the start. Legal referrals contribute almost 30 percent of all admissions in traditional long term communities in the United

Figure IX: Retention curves for residential programs (Bschor, 1986).

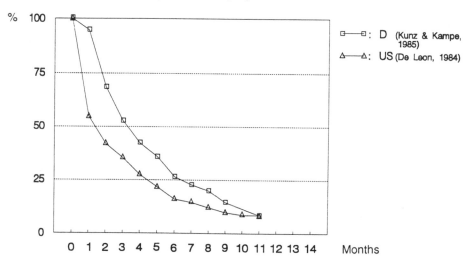

States. However, there has been a decrease during the last decades. For example, over 40% of the admissions to Phoenix House, New York in 1970 were legally referred, compared to less than 20% in 1985 (De Leon, 1987). Most therapeutic community follow-up studies report either small or no differences in post-treatment improvement of legal referrals compared with voluntary admissions.

De Leon (1984) found better outcome for the best success rates (no drug use, no crime) for voluntary clients in Phoenix House studies. However, controlling for criminal background eliminated the significance of the legal referral variable.

Legal referral does not seem to be a significant predictor for successful outcome. The relationship between legal referral and retention is complex. Referrals to a group of therapeutic communities, members of T.C.A. (Therapeutic Communities of America) show that 9 months of retention decreases with age among legal referrals compared to voluntary admissions for whom retention increases with age (De Leon, 1987). This finding is supported by large scale comparisons of retention in therapeutic communities (Pompi and Resnick, 1987). In general, clients referred by the criminal justice system to therapeutic communities, remain longer in treatment than voluntary clients. The relationship between legal referral and treatment outcome is therefore an indirect one (De Leon, 1987). Legal pressure was, however, found to be related to retention only during the first few months that residents are in the program (Pompi and Resnick, 1987; Condelli, 1989).

Most studies do not clearly specify what pressure was being applied by the legal system and they only focus on the relationship between formal legal pressure and resident retention. Many residents however join the programs anticipating court hearings on being arrested and often 'voluntary' residents are under as much pressure from their families and non-legal sources, as 'involuntary' residents are under pressure from the criminal justice system (Condelli, 1989).

Why do clients leave the program?

Hypotheses on clients dropping out, have been made from clinical impressions, such as: the fear of close and intimate relationship with others, not having the support of their family or partners to stay in treatment, feelings of guilt concerning something wrong they did in the community and the fear of this being discovered, an inability to adapt to the way of life in the therapeutic community and unresolved conflicts with staff members. De Leon (1984, 1985) found that long term drop-outs (after 12 months in treatment, reported significantly fewer personal reasons and more program-related reasons for leaving while early drop-outs claim more personal than program factors for leaving.A resident dropping out early may have never really made a choice to abstain from drug use for the rest of his life which is the goal of the treatment in the therapeutic community. He may have sought admission seeking temporary relief from outside pressure. The choice to live a life free from drugs is usually not made before, but after admission to the program; usually only after weeks or months. If this choice has not been made, the resident may leave the program with the clear intention of again using drugs (Kooyman, 1975).

Outcome research of different treatment programs in the United States

As different treatment programs may have different outcome results, different treatment modalities may have different outcomes. Comparison of outcome results, however, produces many difficulties. Program target groups may differ as well as program goals. While the goal of a therapeutic community is a drug-free life independent of a treatment program, the goal of a methadone maintenance program may be substitution of an illegal addiction for a life-long dependency on a legal drug obtained from a treatment program.

In an effort to produce evidence that treatment programs for narcotic addicts work, a large scale research project has been carried out by the Institute of Behavioral Research of the Texas Christian University. Twenty-five agencies participated in the Drug Abuse Reporting Program (DARP) study. The participating programs were classified into the following treatment modality groups: MM (Methadone Maintenance), TC (Therapeutic Communities), DF (outpatient Drug-Free Programs), DT (outpatient Detoxification Programs) and a group consisting of persons that did come for an intake but who were not admitted to a treatment program IO (Intake Only). The results of follow-up studies based on opiate, non-opiate and alcohol use, arrest rates, additional treatment and employment, show a maximum on favorable outcomes for more than half of the sample of the completions and drop-outs (Simpson and Sells, 1981). In a follow-up sample of 2178 persons consisting of black and white males from all treatment modality groups, the following was found: clients in the methadone maintenance program had the largest tenure: 47% stayed more than 300 days compared to 22% of the therapeutic community-clients; 12% of the MM clients had successfully terminated their treatment (5% were still in the program when the DARP-study that started in 1969 ended in 1974) compared to 22% of the therapeutic community-clients. Compositive outcome scores of the one year follow-up were significantly more favorable for the methadone maintenance, therapeutic community and the drug-free outpatient programs than for the detoxification and intake-only groups. The differences remained significant after applying statistical adjustment for client background and baseline variations among the treatment groups (Simpson et al., 1979; Simpson, 1979). However, clients of therapeutic communities that had remained in treatment 90 days or less, showed no difference at follow-up in their outcome results with the detoxification only and the intake only group.

In a long term evaluation study of a sample of the DARP study, it was found that 61% of the sample had achieved abstinence from opiate drugs for a year or longer immediately before the interview. Apart from criminal history, demographic, socio-cultural or drug use history variables were not significantly related to follow-up outcomes. Simpson (1982) reported that 19% of the intake only groups reported immediate and continued abstinence after the initial DARP interview, even though they received no treatment. In general, these persons had better social adjustments than other DARP treatment patients, which were indicated by higher pre-DARP employed rates, lower criminality and less previous treatment. Most had no criminal history and most were self-referrals. These findings are similar to those reported by Waldorf and Biernacki (1981) on 'natural recovery' from opiate addiction. In the DARP study the post treatment outcome showed striking similarities to the subjective evaluation of the treatment-programs by the clients. The highest overall treatment performance evaluation by the clients was given by the therapeutic community-group. This was mainly caused by positive scores on providing insight and understanding and providing help for other problems than drug-related ones. The overall evaluation of the treatment programs by former clients was most favorable in the therapeutic community sample (Simpson and Lloyd, 1979). The therapeutic community group also showed the lowest rate of return to any treatment during three or more years after termination of their treatment, 56% of the TC clients received no treatment compared with 36% to 46% of the MM, DT and IO groups (Savage and Simpson, 1978). Agency differences within the MM and TC program group of the DARP study were not significantly different on post-treatment outcomes for opiate use, non-opiate use, alcohol use, productive activities and criminality (Joe et al., 1983).

Bale (1980) compared the efficacy of three residential therapeutic communities and an out-patient methadone maintenance clinic for veterans in which subjects were randomly assigned to the treatment modalities. Women were excluded from this study because there were too few to provide meaningful comparisons. The nature of the research project was explained to all clients and they were asked for their consent in writing before entering the detoxification unit. The subjects who expressed no interest in further treatment were designated as the 'detoxification only' group. The subjects could only enter the treatment program if they were randomly assigned to a program for one month after admission. After that period they were free to enter any of the other programs. The three different therapeutic communities combined were able to retain 39% of the originally assigned clients; the methadone maintenance program 31%. The compromise '30-day policy' led to only half of the patients in the programs who were originally assigned to them.

In the follow-up study after one year it was found that all groups including the no-treatment group had improved. The MM-clients and the TC clients who had stayed longer than 50 days in treatment used significantly less opiates. The MM-clients did not differ from the no-treatment group in the use of other illegal drugs, while the TC clients used less. The TC clients who stayed less than 50 days did not differ in their drug use from no-treatment groups. The clients of the MM and longer stay TC group were more likely to be employed or attending school and less likely to be in jail or convicted for a serious crime than the no-treatment clients (Bale et al., 1980).

The three therapeutic communities in this study were compared with each other in a two year follow-up. Clients from two of the three therapeutic communities of the study, a professionally staffed community in a hospital and a peer confrontation community staffed by recovered addicts, were found significantly more likely to be working or attending school than the withdrawal-only group and less likely to have been convicted of a crime. The third therapeutic community, an eclectic program employing both professionals and para-professionals, did not differ from the withdrawal-only group on any of the major outcome variables. The two more successful programs, although different in structure and style, were both perceived by their residents as having greater program clarity, order, staff control and orientation to personal problems than the unsuccessful program. Another difference was that the more successful programs involved the residents' relatives, one in evening groups, the other by interviews and orientation while the unsuccessful one did not supply any of these services to the relatives of the clients (Bale et al., 1984).

Short term and long term therapeutic communities

Most therapeutic communities have treatment programs with a planned duration of the time to be spent by the resident in the therapeutic community of 10 to 12 months. Mainly for economic reasons and partly as an effort to attract other persons to in-patient treatment, therapeutic communities with a duration of three months have been founded. Long term therapeutic communities show more favorable outcome results in studies comparing them with short term communities (Aron and Daily, 1974; Coombs, 1981).

What kind of addicts can be successfully treated in short term therapeutic communities is largely unknown and needs further research. A clinical impression is that addicts with a 'junky' life-style and those with a criminal background need long term therapeutic community treatment and that addicts who have another identity apart from that of an addict such as being employed and who have a supportive family, may be successfully treated by short term therapeutic communities. For street addicts admission to a long term therapeutic community seems to be more economical than repeated admissions to a short term program.

Substitution of drug abuse with alcohol abuse after treatment

Alcohol use is commonly regarded as a precursor to opiate use and also as a substitute addiction among recovered opiate addicts. In most therapeutic communities in the re-entry phase of the program, so-called alcohol privileges were given to the clients in treatment. Being able to handle the use of alcohol was seen as part of the rehabilitation. However, many cases occurred in which the residents could not handle the use of alcohol. This led many therapeutic communities to postpone the use of alcohol to the period after graduation, leading to even greater difficulties in experimenting with alcohol without program support. Some programs decided to advocate abstinence from alcohol for life for all residents, including recovered addicts in their staff. Other programs introduced an alcohol-learning period in their re-entry program for those residents who did not have prior alcohol problems.

In the follow-up data of 1409 persons of the DARP study, Simpson and Lloyd (1978) found evidence suggesting a substitution of alcohol for opiate drugs in a small portion of the total sample (less than 10%). Persons who did not return to treatment after DARP termination had lower opiate use, but slightly higher alcohol consumption than those who did re-enter treatment. The use of marijuana and other non-opiate drugs also tended to be associated with higher drinking rates. In another sample Simpson and Lloyd (1981) found evidence for substitution in 13% of the sample. The other findings were confirmed. As heavier drinking after DARP treatment was associated with heavier drinking before DARP treatment, the tendency to misuse alcohol can be interpreted as a return to an old problem rather than an alternative to the use of heroin or other opiates.

To investigate the possibility of a substitution of the drug use by the use of alcohol after treatment, De Leon (1987) analyzed a cohort of the residential population of 1974-1975 on drug change by primary drugs at a two-year follow-up. Of the primary opiate abusers, over 79% were abstinent. There was virtually no non-opiate use. Use of marijuana and/or alcohol had increased significantly, usually to a frequency of 1 to 3 days weekly. The number of non-users of alcohol in this group had decreased from 50.4% pre-treatment to 18.2% post-treatment. The graduates showed the largest difference (73.7% pre-treatment to 5.3% post-treatment). The overall proportion of daily use was, however, low. Daily use pre-treatment/post-treatment was not increased significantly (drop-outs 14.3-11.5%; graduates

5.3 - 10.5%). In the primary alcohol abusers group, over 26% were alcohol abstinent at follow-up. Daily use had dropped from 63.2% to 36.8%. There was relatively little use of drugs (opiates, non-opiates or marijuana) prior to or following treatment. Of the primary non-opiate abusers and the primary marijuana groups, daily use of alcohol dropped significantly (20.7 to 6.9% and 63.2 to 36.8% respectively). In De Leon's study pre-treatment alcohol use was found to be a small but significant predictor of post-treatment alcohol use. Approximately 30% of the successfully treated opiate abusers used alcohol or marijuana more than three times a week for almost one month of follow-up, although virtually no use of other substances were indicated. Their positive status of success was confirmed by the absence of criminality and improved employment. The use of these substances can be seen in a social context.

Regarding the issue of chemical substitution, De Leon's research shows that the primary alcohol and marijuana groups showed little evidence of any shift to opiates or to other drugs. Most opiate abusers achieved abstinence from their primary drug but their use of alcohol increased. The non-opiate abusers (poly-drug abusers) showed an increase in alcohol use with a usual frequency of three times weekly. There were no graduates who reported entry into any drug or alcohol treatment (De Leon, 1987). Thus no support for a substitution hypothesis was found as no shift to serious alcohol abuse could be detected among those who made successful recovery from drug addiction. Difficulties with alcohol may have been acquired some time after the therapeutic community treatment, which was indicated by the significant decrease in non-users of alcohol in the post-treatment period. Use of alcohol may not constitute a new addiction but it signals old negative patterns of coping (De Leon, 1987).

Results of drug-free therapeutic communities for drug addicts are generally more favorable when compared to treatment centers for addicts. Armour (1976) found as a multi-center study that fewer than 25% of the patients had abstained from alcohol for the last half year before the follow-up at 18 months. The more therapy they had the better the results. Also, regular AA attendance indicated the likelihood of disturbance. Costello (1975), in an analysis of 58 studies found that 1 year after treatment 1% of the former alcoholics had died, 53% had a continuing drinking problem and 25% had no drinking problem; 21% could not be traced at follow-up.

Outcome research in Europe

The Lien ward at Dikemark psychiatric Hospital has been one of the institutions in Norway where it was tried to develop a Henderson type (Maxwell Jones) therapeutic community for the treatment of young drug abusers. Vaglum and Fossheim described some results of the evaluation of this program. The plan was to evaluate the treatment of the first 100 young drug abusers. As time progressed three radical changes took place in the program. The different phases called Lien I, II and III, were compared as to the outcome results of the residents admitted and treated in the different phases.

Lien I offered group therapy, no individual therapy and no family therapy. The staff in this period regarded the patients as neglected by their parents and society; there were no strict limits, many splits from the program and a lot of drug abuse during treatment. In Lien II the program was more structured; school and aftercare was added. The staff introduced confrontation groups after they had been to sensitivity training groups. Individual therapy and family therapy were introduced. There were daily community meetings and criminal behavior was confronted in ad hoc groups. When drugs were used immediate discharge followed. However, the patient could come back two weeks later on a contract. The residents in this period started to take responsibility for the program and the control of drug use. In Lien III many of the staff left for various reasons. The environment became less confronting. More individual therapy and concern and support from the staff for psychotic and near psychotic patients was present than in the other phases. Most patients received individual therapy and family therapy. A fourth group in the research consisted of clients that received other than therapeutic community treatment. The clients of the sample that were interviewed in the follow-up study (96%) showed no statistically significant difference in background, except for low social economic status being more present in the Lien II subgroup than in the others.

In the last year, before the follow-up, no drug abuse was found in 41% of the Lien I group, 63% in Lien II, 34% in Lien III and 38% in the group not treated in Lien. In the Lien II group only 15% had used drugs or died in the year before the follow-up while in the other groups around 30% had been using heavily. When compared, the difference in the results of those who were injecting drugs before admission is still more outstanding: 70% stopped the abuse in the Lien II groups, while this was only around 30% in the other three groups. Patients who had used psychedelic drugs, however, had improved least in the Lien II group and had the best results in the Lien III group. Thus, different groups of drug abusers got the best help from partly different therapeutic programs. Psychedelic abusers who were often near psychotic or psychotic, did best in a supportive limit-setting environment combined with individual psycho-dynamic therapy and family therapy. The non-psychotic abusers of opiates and central stimulants were better helped by a more confronting environment where drug abusers themselves took the responsibility of keeping drugs out of the program. Many of them were given individual and family therapy. In both groups successful treatment seemed to be facilitated when the patient managed to get into a relatively safe relationship with therapists, friends and family. As the researchers stated:

many of the residents said: „Drugs, that's what you need when you have a lack of people (Vaglum and Fossheim, 1980)."

The first therapeutic communities based on the self-help philosophy of the American therapeutic communities were founded in England. Ogborne and Malotte conducted a research of the first 100 residents admitted to the concept house founded in London in 1970 (Ogborne and Malotte, 1977). In the follow-up study 87 personal interviews were obtained from these 100 residents at least 6 months after they left the program. Less than 10% had completed the whole program; 30% of all admitted persons showed no or only sporadic drug use. This was positively related to employment and lack of criminal convictions. There were 17% who abstained from the use of psycho-active drugs, except alcohol (this was confirmed by urine-tests). They had all stayed over a month while regular injectors were least likely to have stayed a month; 9 out of 11 of the abstainers had completed the program. Of those who abstained, 40% had experienced some alcohol problems. The abstainers and sporadic users had some educational advantages over the other persons in the sample. The positive effects on drug use were found to be highly related to length of stay in the program and having completed the program or not.

In a more recent research in the program of Phoenix House, London, changes in self-esteem during the stay in the therapeutic community were measured. Negative feelings about the self are regarded as being an important risk factor for drug abuse. A gain of 94% was found in one of three therapeutic communities, compared with 25% in the two other therapeutic communities of the program. The therapeutic community with a more favorable outcome had higher retention rates, a more stable program, less encounter group meetings and more time devoted to practical activities and learning social skills (Kaye, 1987). These results support the findings of Biase in Daytop Village New York, that the availability of education during the stay in the therapeutic community is related to an increase in self-esteem (Biase, 1986).

Wilson and Mandelbroke compared the conviction rates of the first 61 residents admitted to the Ley Community in Oxford during the two years before and the two years after discharge. The residents were divided into three groups; a short-stay group (less than 1 month), a medium-stay group (between 1 and 6 months) and a long-stay group (6 months and longer). The three groups did not differ on pre-admission background characteristics. More than 80% of each group had a history of opiate abuse. The conviction of the long-stay group had dropped from 60% to 10%, from the medium-stay group from 70% to 40% and the rates of the short-stay group had not changed but remained 57% (Wilson and Mandelbroke, 1978).

Wilson and Mandelbroke repeated the research 10 years after discharge. Sixty of the 61 persons were traced; six had died; four of the short-stay group had died due to drug abuse, one of the medium-stay group of natural causes and one of the long-stay group as a result of suicide. Only three of the long-stay group of 20 persons were re-convicted during the 10 year period (15%), compared with 70% in the medium-stay group and 85% in the short-stay group. The findings were extremely promising when compared with those of a ten year follow-up of patients of a London drug clinic (83% re-convicted) (Gordon, 1973). In a follow-up interview 90% of the same group of clients could be included. The interviews were carried out between 2 and 4 years after leaving the community. Relapse into regular or irregular drug-injection was found to be correlated with the length of stay in the program (Wilson, 1978).

Deissler published in 1981 evaluation results from Aebihus, a therapeutic community in Switzerland based on the concepts of Synanon. During an observation period of 6 years, 54% of the residents left prematurely. Eighty of 100 Graduates could be interviewed 1 month to 5 years after leaving the program. Only 6% were total abstainers, 21% had relapsed into opiate use; the others had sporadically or regularly used alcohol and soft drugs, such as cannabis. However, 70% earned their own living, 85% had a more or less regular job, while 20% had been engaged again in some criminal activity (Deissler, 1981, referred to by Uchtenhagen, 1985). Also in Switzerland Bernath studied a program where non-paid non-professional staff, partly with their family, lived with a small group of addicts in a community of about 25 persons in total. Seventeen of the 25 former opiate addicts where contacted after they left their community. Although 70% had relapsed one or more times, 47% had been without the use of drugs for 3 years. The more advanced the persons had been in the stages of the program when they had left, the shorter had been their periods of relapse (Bernath, 1978, referred to by Uchtenhagen, 1985). In the program of Daytop Germany, Lutterjohann found a correlation of completion of the program and self-reported abstinence. He also indicated the usefulness of Rational Emotive Therapy in therapeutic communities to prevent relapse (Lutterjohann, 1984).

In Italy the first 203 clients admitted between 1975 and 1984 in the therapeutic community Casa Verde in Milano were included in a follow-up study (Gori et al., 1984). The program had shown three stages of development in the 10 years of the study, during which the time between first contact and admission to the therapeutic community extended from 2 weeks in the initial stage of the program to about 6 months in the later stage. In the last stage outpatient psychotherapy groups were part of the program before admission. With the increasing time before admission, the early drop-out-rate (leaving before 1 month) in the therapeutic community was reduced from 40-50% to 4%. The graduate percentage had risen to 50%. There was no significant change in the retention of those leaving the program prematurely between 1 month and completion. Of the drop-outs, 10% could not be traced in the follow-up study within a month, 17% could be regarded as fully recovered when considering such factors as : heroin use, prison detention, re-entering the therapeutic community for treatment, compared to 34% of those who left between 1 month and graduation and 70% of the graduates (Capitanio, c.s., 1985).

In a study of residents in the Phoenix House therapeutic community in Oslo, Norway 144 residents were followed during their stay in the program. Thirty percent remained longer than one year and 20% completed the total program. Significant differences were found between drop-outs and completers. Among those who completed the program, there were more males, more persons who were using amphetamines in high frequency and alcohol in low frequency before admission. They also had less schizotypal traits than the drop-outs (Ravndal and Vaglum, 1991). In a separate study a representative sample of female addicts was followed in the Phoenix House program and two months after the program, which all but one had completed. All women in the no- success group had entered into destructive relationships with male co-residents, while none of the successful women did so. A repetition of relationships to parents was repeated in partner and peer relationships, which were strongly related to outcome (Ravndal and Vaglum, 1992). In another study 15 HIV-positive clients of Phoenix House, Oslo were followed with a follow-up four years after the start of the treatment. Half of the HIV-positive residents completed the inpatient treatment compared to only 27% of the HIV-negative residents. However the completion rate of the total program including the outpatient re-entry phase was quite similar to that of the HIV-negative clients (20% against 19%). Of the HIV-positive clients at the four year follow-up 3 had died, of the remaining 13 only one person was not using drugs, the others had relapsed, two had been diagnosed as having AIDS. There had been no relationship between time in the program and success (Ravndal,1992). This study, although the numbers are small, questions the desirability of having separate treatment programs for HIV-positive addicts which is more directed to their needs. An extensive research project, called the Swedate Project was started in Sweden in 1988. It is a combination of process and outcome evaluation; 31 different programs were involved, including 8 therapeutic communities for youth and 16 therapeutic communities for adults. Complete intake data were collected of 1164 clients, from which 570 randomly selected clients were included in a follow-up study. There was a close cooperation of researchers and treatment staff of the different programs during the research project. To be able to describe the different philosophies and goals of the programs, not only the directors were interviewed but all treatment personnel.

The therapeutic communities were mainly of two types:

- Those programs based on the environmental therapy of Maxwell Jones, with a democratic non-hierarchical structure, where staff and clients share the responsibilities for the community. Addiction was seen as a symptom of psychological and social disturbance and psychotherapy was the main tool to solve the addiction problem.
- The other type developed as the 'Hassela pedagogics', communities inspired by the Russian educator Makarenko. Addicts were seen as unsocialized youth exhibiting norms and values destructive to their surrounding society. The treatment is based on re-education using traditional norms of a healthy family. These programs were inspired by a socialistic political ideology.

Some of the findings of the intakes are worth mentioning here: 61% of the addicts came from broken homes (78% of the programs for youth), 22% of the fathers had alcohol problems, 22% had no contact with the mother, 45% had no contact with the father. Of the drug-addicts 39% used mainly cannabis. The dominant drugs for the others were amphetamine: 46%, opiates: 18%, other drugs: 6%. For adults only the percentages were: cannabis: 26%, amphetamines: 47% and opiates: 22%. Of the selected 560 clients in the follow-up study, 448 were interviewed, 112 were missing of which 11 were dead. The response rate was 80%. They were interviewed 12 months after they had left the program. Thirty-five percent of the youth and 36% of the adults had never used drugs. Another 12% of the youth and 16% of the adults had used drugs but were drug-free during the last 6 months prior to the interview. Many of the persons interviewed had been drinking heavily. Of all those who were drug-free in the last 6 months before the interview, 5% of the youth and 16% of the adults had been drinking more than 40 grams of alcohol daily.

The different therapeutic community programs were divided into clusters with similar treatment philosophies and programs. The programs with no unity between the philosophy and the actual practice, or with conflicting ideas among staff, were put into one of these clusters. This last cluster showed the poorest overall results. However, it is interesting that among the more homogeneous clusters there was a vast difference found in outcome results ranging, for instance, from 64 to 0 in the cluster of therapeutic programs for adults with an environmentally therapeutic approach, based on humanistic philosophy. Most of these units used T.A. (transactional analysis) as the main therapeutic method (Stensmo, 1988; Björling, 1986, 1989; Segraeus, 1986).

The research of the Swedate project found a correlation between time in treatment and positive outcome. Of all clients in the follow-up study, 51% reported no use of drugs one year after treatment. Women did somewhat better than men. Some, however, had a pattern of heavy drinking or frequent use of benzodiazepines. With very strict criteria (no drugs, little alcohol and no benzodiazepines) there were 37% successful cases. Only 10% had no drugs problem, no forced institutional care, was socially integrated and had no need of authorized social support or welfare after treatment. The persons in this most successful group were compared with those in the worst group, scoring negative on all success criteria. On the background criteria only family factors in early chidhood were found to be significantly different. The worst group had more early separations from their parents and admissions to children's homes. In the worst group there had been more parents with alcohol or other substance abuse and mental problems. Of the worst group 93% was unemployed at intake against 54% of the best group. Of the best group in 57% of the cases the discharge was planned, of the worst group only 8%. In view of the total sample the staff's view proved to be a good predictor for treatment outcome. A good treatment climate within the staff was also found to be related with positive outcome results (Segraeus, 1992). There was a negative correlation between successful outcome and the percentage of former drug abusers among the staff. The researchers also concluded that an empirical study of matching clients is almost impossible due to the complex interaction within and between clients in treatment (Berglund et al., 1991).

In a study of the first 40 residents of the Therapeutic Community De Kiem in Belgium, 33 were contacted one year after admission, 11 were still in treatment and 22 had left. Of the latter, 38% had again committed a suicidal attempt, 50% had been admitted to a psychiatric hospital and 25% had problems with the police. However, a quarter of the splittees had remained drug-free (Broekaert, 1981). During the first 6 years of the Therapeutic Community De Sleutel in Belgium 70% of the residents who had been admitted had left within the first 4 months of stay (Maertens, 1982).

In a research in a similar Belgium therapeutic community in Brussels, De Spiegel, the following was found: Residents who had stayed at least 30 weeks showed a clear decrease on legal and illegal drug use, a decrease on prison detention and admission to a residential center for treatment. This positive outcome was related to being employed. Of the 128 residents on the study, 17% were still in the program after 52 weeks. The male\female ratio was 77/23%, 53% had mainly abused opiates, 30% alcohol, 16% amphetamines or other drugs. The persons in this group started using at an average of 4 years before admission; 60% were interviewed in the follow-up study: The non-respondents were not statistically different. One year after leaving the program, 19% had not used any drugs. There was no difference between the early drop-outs (less than 8 weeks), the medium stayers (between 9 weeks and 30 weeks) and the long stayers (longer than 30 weeks). However, when ex-residents relapsed either into the use of alcohol or soft or hard drugs, the long stayers were using for shorter periods, while early drop-outs either stopped or relapsed and continued to use (Vandenbroele, et al., 1989). After a crisis detoxification center was opened, the early drop-outs' rate of the therapeutic community De Spiegel dropped both for residents admitted to the therapeutic community from his detoxification center, and for residents admitted directly to the therapeutic community (Vandenbroele, 1991).

In a pilot study in the Emiliehoeve therapeutic community in the Netherlands, subject of this current research, Kooyman found a sharp increase in the retention potential after the program had changed from a democratic therapeutic community into a clearly structured program (Kooyman, 1975a). He also described the promising outcome results of a population present at one particular day in the Emiliehoeve therapeutic community; 50% of the 23 residents had not relapsed into drug abuse or criminal behavior and neither had they been admitted to a psychiatric hospital at their follow-up interview (Kooyman, 1985b).

In a study of the drug addicts admitted to the therapeutic community Breegweestee, one of the therapeutic communities of the Nieuw Hoog Hullen Foundation, modelled after the Emiliehoeve, no significant difference in social background variables and drug addiction history was found between the group that was admitted and the group treated in an ambulatory program. This finding contradicts assumptions that therapeutic communities treat a highly selected population (Van der Velde and Jongsma, 1985). In the same foundation a therapeutic community for alcoholics was modelled after the hierarchical therapeutic community Breegweestee, which originally admitted only drug-addicts. A research project proved that this hierarchical model was also effective for alcoholics. A follow-up study carried out among clients of the New Hoog Foundation included 881 (638 of them alcoholics) clients all admitted to the detoxification clinic. From there they were referred to one of the two therapeutic communities, to a short-stay in-patient program, to an ambulatory treatment program or to other in-patient treatment elsewhere. Follow-up information was gathered 8 months, 18 months and 30 months after the intake using a questionnaire that had to be returned. The second follow-up showed a 53% response: 69% had no alcohol abuse at the time of this follow-up. Positive outcome was related to low psychiatric scores at admission, higher social economic class and lower egocentricity. For drug-addicts serious criminality was related to poorer outcome. Ex-residents of both therapeutic communities showed the greatest improvement, especially among those who had graduated from the programs. Compared with the ambulatory treated patients, the residents of the therapeutic communities had shown more problems at admission. The poorest results were found in the group that had received other in-patient treatment than in one of the two therapeutic communities (Van der Velde et al., 1989). The most important treatment variable related to successful outcome was time spent in treatment (Schaap, 1987).

In the Jellinek Center the effect of program changes was studied on retention. Two therapeutic communities (Parkweg Binnen and Buiten) had similar programs up to 1983. The Therapeutic Community Parkweg Binnen shortened the program time in the therapeutic community from 9 to 5 months. The other Therapeutic Community Parkweg Buiten kept the duration to 9 months. Also from 1983, network treatment was started with the persons on the waiting list (at that time 3 to 6 months) including group meetings of the addicts and parent groups. The parent groups proceeded though different phases parallel to the treatment of the client. The clients who left the therapeutic community were divided into two groups: a drop-out and a success group, based on time spent in the therapeutic community and on therapists' ratings. There was no follow-up to evaluate the success after the residents had left the program. In the success group there were statistically more males, more persons with a higher education, higher professional level, more persons referred from other treatment institutions, persons with longer duration of the addiction and persons with less abuse of pharmaceutical drugs. There were more persons who were not staying with their parents after leaving the program among the drop-outs (32%) than among the success group (11%). Of persons staying in the therapeutic communities longer than one month, 75% of the success group had parents attending at least one parent group against 58% of the drop-out group. Residents with parents attending parent groups were divided in short- stayers (less than one month) and long-stayers (long than one month). The percentages were 58% and 77% respectively. Before parent groups were introduced the percentage of short-stayers and long-stayers among all residents had been 49% and 66%. These differences were statistically significant. In this pilot study a reduction of drop-out after the introduction of parent groups in the period in which the clients were seen before entering the therapeutic community, could be demonstrated (Nabitz and Hermanides, 1986).

CHAPTER 10

The Emiliehoeve story:
From chaos to structure

When the Emiliehoeve was founded on February 14th 1972, there was a need in the city of The Hague in the Netherlands for a center where drug addicts who wished to stop their addiction could be treated. There was a small methadone maintenance program for opiate addicts an the out-patient clinic, however, in that period the majority of the addicts used amphetamines as their main drug. Heroin was not then available on the drug scene. The opiate addicts used opium obtained mainly through the Chinese community in Amsterdam. For decades this community has maintained a small group of opium users. Only after the Amsterdam police had successfully repressed the Chinese opium sellers, did other dealers bring heroin on to the Dutch drug market (Kooyman, 1984).

When amphetamine users became psychotic, they were frequently admitted to a psychiatric hospital. In these hospitals they usually learned to combine their drugs with pharmaceutical drugs prescribed to them or borrowed from fellow in-patients. The out-patient treatment programs of the addiction centers as well as the treatment in in-patient drug clinics (i.e., the Parkweg Clinic of the Jellinek Center in Amsterdam and the Essenlaan Clinic of The Bouman Foundation in Rotterdam) or the treatment in psychiatric hospitals, were not successful in treating the addiction. Due to a lack of structure and strict rules, drug use continued after admission to these medical model clinics.

As none of the existing treatment possibilities were successful in treating the addiction to drugs, a need was felt by the medical community to establish a treatment program as an alternative to a methadone maintenance program, rotating admission to psychiatric hospitals or medical model clinics. As a response to this need, the Emiliehoeve was founded as a therapeutic community for addicts where the clients could learn to live a life without any drugs. No methadone would be given and tranquillizers and sleeping medication would only be given for a limited period. After an exchange of land with the city of The Hague, the community was established on a farm that had become part of the grounds of the psychiatric hospital `Bloemendaal'. Initially there were only 10 places for male residents.

The model for the therapeutic community was drawn from therapeutic communities for psychiatric patients, set up in accordance with the ideas of Maxwell Jones (Jones, 1953). The responsibility was to be shared by staff and residents and the goal was set on re-socialization by social learning through social interaction. Although the Emiliehoeve was located within the premises of a psychiatric hospital, the staff had departed from the medical model. It soon became evident that the democratic system did not function for the population admitted in the therapeutic community. The community became in fact anti-therapeutic. Negative elements among the residents took over control. Residents were leaving prematurely and thus the last remaining residents had to be discharged because they had relapsed into drug use.

The system was then changed introducing more and more structure. Concepts and therapeutic tools from the American self-help therapeutic communities for addicts were introduced. Subsequently the American model with a hierarchical staff and residents structure was adopted. Finally, after a period in which this model had been applied in a very rigid way, this hierarchical model was modified and applied with more flexibility.

The development of the therapeutic community differentiated into 8 distinct phases:

I.   The a-therapeutic community phase (February 1972 - June 1972)
II.   The confrontation and intimacy phase (June 1972 - September 1973)
III. The hierarchical adoption phase (September 1973 - January 1975)
IV.   The post-professional phase (January 1975 - June 1976)
V.   The closed community phase (June 1976 - April 1977)
VI.   The integration phase (April 1977 - May 1980)
VII. The open program phase (May 1980 - September 1981)
VIII. The adult program phase (September 1981 - the 1990s)

Phase I: The a-therapeutic community phase (February 14th 1972 - June 13th 1972)

On the day of the opening of the therapeutic community, the psychiatrist responsible for the program had invited all workers in the field of addiction in the Netherlands known at that time. He told the sceptical audience that the Emiliehoeve would be the first center in the Netherlands where addicts would be successfully treated. This statement of course led to some cynical remarks from the audience. He had invited a colleague who had recently visited a therapeutic community in New York to give a presentation on the application of encounter groups in the American therapeutic communities for addicts. That same day the first addicts were admitted from a ward of the psychiatric hospital.

The main therapeutic tool was considered to be group therapy. The groups in the Emiliehoeve however, were initially led mainly on analytical lines, due to a lack of experience of the staff in other group therapeutic techniques. Apart from the psychiatrist, there were two other therapists, a psychologist, and his wife who had not yet finished her studies in psychology. There were two group leaders (socio-therapists), one part-time creative therapist and one part-time supervisor of the farm and garden activities. There were no nurses in the team. This was a revolution to still be part of the organization of a psychiatric hospital. The Emiliehoeve staff thought that 'nursing' was the last thing addicts would need. In the first years nurses from the nearby psychiatric hospital slept in the therapeutic community at night. From the third phase onwards staff no longer slept in the therapeutic community. The night and weekend duties were then taken over by residents working in pairs on a rotating schedule. A nurse from the psychiatric hospital had the task of a liaison person with the hospital management.

The residents of the first phase were admitted usually after a stay of one or two weeks in the psychiatric hospital where detoxification took place. In the night of the opening day the first residents arrived. The first residents were addicted to amphetamines and/or opiates. In a later stage, from Phase 2 onwards, alcoholics were also admitted if they were of the same age group (between 16 and 40 years).

At the onset, after discussions with the residents, the following rules were agreed on:

a.   No use of drugs including soft drugs such as hashish.
b.   No leaving of the premises without permission from the staff.
c.   No smoking in the sleeping rooms.
d.   The residents should clean the house themselves.
e.   There would be urine controls taken once a week on randomly selected days.
f.   Visitors could come only on Wednesday evening or Sunday afternoon.
g.   The residents should take initiatives in planning the work program at the farm.
h.   Participation in the therapy groups and other group meetings was mandatory.

The initial group of residents consisted of ten young male drug addicts who had a history of multiple drug use with amphetamines being the main problem. Group therapy sessions and community meetings were the only obligatory activities, the former being held three times a week. Working in the garden and in the house was encouraged by the staff. Some time was spent on recreational activities such as painting, woodcarving and sports under the guidance of a part-time staff member. All plans were discussed by the entire community. Decisions were taken collectively by the staff and patients in a democratic way. All residents, regardless of their status and time in the program, were allowed one vote. A simple majority vote would automatically implement policy. There were 10 residents and 5 staff of whom 2 were part-time. The residents always retained the votes to overrule any staff suggestion. Complicating the process was the fact that rarely could the entire staff attend the community meeting where daily decisions were determined. The explicit understanding was that 'treatment' would become a mutually shared responsibility. In so doing, the Emiliehoeve program attempted to eliminate the 'we'-'they' dichotomy between patients and staff. The implicit assumption, which was later proven incorrect, was that addicts during the initial phase of their treatment, when given the opportunity, would take reasonable, realistic and rational decisions. The staff was instructed to reinforce all positive and productive behavior by verbal compliments. Concurrently, the staff ignored any self-destructive and anti-social behavior, because they assumed disruptions would abate on receiving no special attention. Addicts, during the early phase of their treatment, did not respond to this permissive psychotherapeutic orientation, because they were continually testing limits.

The Emiliehoeve experience indicates, that unless there is direct therapeutic intervention and consistent limit-setting by the staff, that disruptive and dysfunctional behavior will continue to escalate. For example, when one angry resident deliberately threw a glass of water on the floor, the staff remained silent. Receiving no therapeutic restraint, the resident became more destructive by not only breaking glasses, but also throwing a chair through a window. In as much as the resident exceeded all acceptable limits, he placed himself in a 'no win' situation. The staff was polarized into a reactionary position of having no other option but to expel him. In retrospect, tragically, neither the staff nor the other residents intervened in this and previous similar episodes, so what started as relatively minor acting-out, resulted in expulsion.

The staff's attempts to reinforce positive and productive behavior proved ineffective; such action had only temporary effect. When residents were cooperative and were congratulated for their contribution one day, they regressed the next day. Whether this failure was caused by the treatment-model or because the resident, who was most respected by his peers, engaged in defiant and rebellious behavior, cannot be determined. By accepting the resident leader's negative behavior, the staff inadvertently indicated an inability to manage him. He, thus, became a powerful noxious force. The staff felt frustrated and demoralized because they were unable to control and contain this disruption.

Significantly, during this period of turmoil, suicide attempts by residents occurred as did premature departures. This was probably due to a lack of structure and a lack of effective therapeutic input. Still, utilizing the Jones Model, the next day's agenda was determined at the daily evening community meeting attended by program participants and professionals. The next day, however, the residents did not do what they had voluntarily agreed upon less than 24 hours before.

When it became evident that the egalitarian Jones Model was less than effective, the staff decided to augment it with a token economy. Residents could be either rewarded or punished by receiving points which would be translated into pocket-money at the end of the week. With positive behavior, points could be earned. Negative behavior resulted in loss of points. This token economy system was linked to the amount of pocket money and weekend privileges. This token economy system helped to improve the situation. The token economy enabled the staff to retain a psycho-analytic orientation, to remain relatively passive, and to be interpretive. Group therapy was conducted in the traditional psycho-analytic way (Adams, 1978; Wolf and Schwartz, 1962). The therapeutic focus remained with past relationships and events. Interpretations of previous conflicts were provided. Rarely did the staff attempt to connect the past with the present. Conflicts were discussed in an intellectualized manner. The thrust of therapy was to produce insight rather than to attempt to influence behavioral change. Although the program participants were called residents, the staff was unaware that they treated them as patients.

After four months, the remaining two residents were discharged because they continually abused drugs. In view of the irrefutable evidence that none of the original residents were helped, the staff considered other treatment philosophies. Using drugs in the comfortable setting of the farm had become a mere substitute for using drugs in the streets. One positive outcome for the staff was a definite lesson in 'what not to do'; it helped the staff to get a clearer idea of how a treatment center for addicts could run better.

After these first four months the staff concluded the following:

a. Motives for seeking admission to the center could be others than the desire to stop using drugs. For instance, to stay out of prison, to be able to pay off debts out of social allowances still received after admission, to become clean in order to start a cheaper habit, or to continue using drugs with a holiday in the country.

b.   Residents not admitted on a voluntary basis continued using drugs and proved to have an adverse influence on the group. On occasions, when one member of the group went out to buy drugs, nearly all the others relapsed into drug use upon his return.

c.   When the group had not been participating in daily activities, the tendency to take drugs was far greater than on work days.

d.   The program on the farm had to be more strictly structured. After two months a token economy system was introduced in order to stimulate activities on the farm, and it was at this point that relapses in to occasional drug use decreased.

e.   Cohesion within the group appeared to be necessary in order to obtain worthwhile results from the therapy program.

f.   In order to appraise the success of this form of treatment, a follow-up study is necessary.

For the next month, the program was in transition.

Phase II: The confrontation and intimacy phase (June 13, 1972 - September 9, 1973)

It was decided not to admit residents only because they had expressed their need for treatment. The interview to test the motivation of each resident was carried out by all staff members on duty, in this phase, and no longer by the psychiatrist only. Two of the residents that had been admitted in phase I were re-interviewed and re-admitted.

Contrary to Phase I, in Phase II women were also admitted. This had some positive effects, such as a more home-like appearance of the living environment. It soon also led to some problems. The first woman admitted, agreed that she should stop her habit of sleeping with various men. However, she resumed her habit within a few days. This led to discussions in the group. It became clear that the men tended to boast of their success with the woman and that the woman made love with men to test the reactions of other men, whom they really liked better. It was clear that sex used in this way was destructive and had become a narcistic manipulative power game. As a result, the residents and staff proclaimed a no-sex rule during the stay in the community among residents. Residents began to relate to each other as brothers and sisters. The group became transformed into a family where there was mutual sharing and concern.

In the first months of this phase an important change in the program took place. The psychiatrist and the group therapists participated in a marathon-weekend led by a former staff member of Phoenix House, New York. He was a recovered addict and had been the first director of Phoenix House, London. This was their first experience with encounter groups. The weekend was from July 7-9th and lasted 54 hours, interrupted with only four hours of sleep

(sleep-deprivation was used to reduce resistance, a common practice in marathon-therapy groups at that time). One week later the staff introduced encounter groups into the community after the group leader had paid a visit to the Emiliehoeve farm. In the encounter groups the residents not only learned to express themselves but also to cope with their emotions. Participants were helped to understand that many of their current feelings and much of their behavior was influenced by unresolved conflicts from their past. Tranquillizers and sleeping pills which had been given to residents during the first weeks of their admission to the community were no longer given.

The impact of this introduction of encounter groups during which residents as well as staff members confronted each other with their behavior in a direct emotional way was enormous. In the encounter groups the groups usually started with topics from the 'here and now'. Often by the screaming out of emotions, conflicts from the past surfaced up and could be worked through emotionally. Other elements were introduced into the program. After the psychiatrist and one of the group therapists participated in marathon groups led by Daniel Casriel on two consecutive weekends in August 1972, the scream and bonding techniques of Casriel's New Identity Process were included in the encounter group techniques (Casriel, 1972). Hugging between staff and residents and among residents occurred in these groups. The participants learned to overcome their fear of intimacy, of physical and emotional closeness.

After new residents had successfully passed the interview, a ritual followed: they were thrown into the canal in front of the farm and were given overalls to wear. After the first month they were given their own clothes to wear; junky or flower-power outfits were no longer acceptable.

Apart from changes in the therapy groups, other new elements were included in the program, such as speaking engagements, whereby one resident introduced a topic to the group, which was then followed by a discussion. Residents' visitors could only come during weekdays, and no longer on Wednesday evenings. Visitors to the program, such as persons involved with organizations who worked with addicts, were welcomed in groups. Older residents were stimulated to take part in courses or evening classes outside the Emiliehoeve. All staff were sent to encounter weekend-workshops, organized by centers of the human potential movement.

Visits of relatives and friends and weekend leave to visit relatives were postponed from one to ultimately four months after admission. At the first weekend leave, the resident was accompanied by an older fellow-resident. Confrontations with relatives at earlier periods had frequently led to strong emotional reactions and often to premature leaving of the program.

Drug use in the house had stopped completely in this second phase. An important event was the recovery by residents and staff of a church-organ from the house of a dealer, who had removed the organ some months earlier with the help of some residents from the farm providing them with drugs, and its triumphant return on the roof-rack of the psychiatrist's car (with a staff member playing: „We shall overcome" on it). This happened on August 17th, 1972. Since then, no drugs have been used in the Emiliehoeve. All randomly taken urine samples remained negative.

Residents on the farm were given more responsibility. They started to cook their own meals. From December 1972, a coordinator in charge was elected jointly by the staff and residents, and made responsible for the delegation of all household tasks and activities, such as biologically dynamic gardening and furniture-making. Heads for these projects were selected in the same way. From March 1973 a resident was appointed responsible for the administrative tasks in the community, such as the supervision of payment for shopping and pocket money. Although there was a positive effect after the residents had elected a coordinator, some residents used the democratic system as a device for escaping responsibility. They would for example choose a coordinator from whom they could expect very few demands. It was noticed that decisions taken jointly by staff and residents in a democratic way, could prove to be anti-therapeutic, for instance in allocating the work, approving weekend leave or keeping pets. When work assignment was largely left to the decision of residents, it led to the formation of sub-groups on the farm. Residents appeared to hesitate to cancel weekend leave requests from their fellow residents, since their own requests might be similarly treated on another occasion.

The token economy system was maintained during this phase and linked with pocket money. After a proposal by the residents, all allowances from various security funds or other income, was contributed to a common pot, from which the residents received pocket money, money for clothing and recreation activities. It was found unfair that residents could have great differences in allowances. From the pot, debts accumulated before admission could be partly paid off. During the first five weeks residents did not receive any pocket money.

The system of taking decisions democratically by all persons in the therapeutic community by consensus, such as the choice of a coordinator or dividing the work, was still used. However, it resulted often in avoiding problems by choosing a nice person as the coordinator; a person who could easily be manipulated. Often subgroups of negative persons were formed and clear job responsibilities were avoided. Being the coordinator was not a position residents volunteered for. They volunteered for jobs they were already good at. Therefore, a lot of possibilities for learning to use the democratic decision process were avoided. In making decisions the staff was always in the minority.

In this second phase, parent groups were started. Parents of residents who had been in the house for at least two months, were invited to attend bi-weekly meetings. In these meetings, led by staff members, parents could discuss as a group their contribution to problem. It was found that many parents had feelings of guilt and therefore had usually done too much for their children, such as giving money, or paying debts, which had only resulted in the continuation of the drug abuse. Parents were told about the objectives and goals of the program and encouraged to lead their children back to the therapeutic community in the event of them leaving prematurely.

This second phase can be characterized by strong personal involvement by the staff, who also made the residents feel like pioneers and responsible for the program. There was an emphasis on learning to handle emotions and close physical contact. The philosophy of the American self-help programs had been implicitly adopted. Although decisions were taken by the whole community, residents tended to decline responsibility for the daily activities on the farm. It was felt that more pressure was required for residents to learn to accept responsibility and adapt their attitudes, behavior and life-style in order for them to cope with stress and the demands of society after discharge from the Emiliehoeve. This phase ended with a workshop for the staff on the basics of the American therapeutic communities, led by the group leader who had run the earlier mentioned encounter marathon in May 1972.

Phase III: The hierarchical adoption phase (September 9, 1973 - January 14, 1975)

In the workshop, preceding the new phase, the staff was trained in using a hierarchical staff and residents structure as a therapeutic tool. During this workshop the staff was trained to set up general meetings in the house and to give structured verbal reprimands (so-called 'haircuts'). At the end of this workshop the Emiliehoeve staff discussed for 14 hours the decisions about whether or not to choose for a hierarchical structure. This choice was not easy. Especially because in those days horizontal democratic structures were replacing hierarchical ones in social work and health care organizations in the Netherlands. In the middle of the night of 9th September, the residents were called out of their beds for a general meeting in which the new staff and residents structure was announced. To the staff's surprise the residents did not rebel against this decision, which was taken without asking their opinion. The staff had made it explicit that there were two groups of people in the community: residents that came for help and staff who were paid to work in the program. From then on decisions were no longer all taken by staff and residents together.

So-called clinical decisions, such as the position of the residents in the work structure, privileges, weekend leaves, admission and discharge, were taken by the staff only. Decisions concerning the daily activities were delegated to the residents through the work structure. The token economy system was abolished as it turned out to be a means of avoiding direct confrontation of staff members and residents, in disregard of the philosophy of the encounter groups. Residents were expected to do their jobs for one hundred percent. To the cardinal rules of no drugs and no alcohol, rules of no violence and no sex with fellow residents were added.

In August 1973, a separate foundation called 'Maretak' was set up to provide housing for residents after they had left the therapeutic community. It was felt that a halfway house was necessary for a certain period after having left the safe atmosphere of the therapeutic community. In February the first residents entered this re-entry house. There was not yet much of a re-entry program then. A part-time social worker had individual meetings with the first re-entry residents. The re-entry residents were stimulated to take up education in this period. Most of them worked as staff aides in the program for several months.

On January 15th 1974, an ambulatory induction center was opened in the city of The Hague. Before admission to the therapeutic community, new clients had to come to this ambulatory induction center daily, for one to two weeks. Re-entry residents and older residents from the therapeutic community were involved in running interviews and group therapy with the inductees. By asking persons who wanted to come to the therapeutic community to perform certain tasks, for example to change their clothing or to write their life- story, it appears that admission to the Emiliehoeve became more attractive to them.

By building extra bedrooms in the old barn of the farm, the number of residents in the Emiliehoeve was able to be extended from ten to twenty. An interesting phenomenon occurred, that was also seen later in other expanding therapeutic communities (i.e. in Sonnenbühle in Switzerland). When the number of residents in the house grew above 15, usually some residents left until the number of 11 or 12 was reached. At that time all group meetings were held with all residents together. After this was changed into two encounter groups running simultaneously when the number reached 16 residents again, this phenomenon stopped and the community was able to expand to 20 residents. Also the encounter groups became more dynamic when both groups had other participants. Separate groups for new residents and groups for female residents were also started in this phase.

The work program with a hierarchical structure provided an opportunity for the residents to explore, accept or even seek more responsibility for themselves and others. In addition, the work program created stress situations for the residents. These problems and difficulties were discussed during the encounter groups. In these encounter groups all participants were equally important. Also the staff who were running the groups could be confronted.

Each day, a morning meeting run by residents was held followed by department meetings of the various work departments. The staff had to learn to present themselves as figures of authority. Also within the staff there were now more clearly defined responsibilities. The psychiatrist, who was already assumed to be the head of the program by persons outside the therapeutic community, formerly assumed the title of Director.

One staff member had started parent groups at the end of the previous phase, which became bi-weekly meetings of parents of residents. Parents of residents were approached during the second month of the resident's stay. Usually a home visit followed and the first attendance of a parent to the group was usually planned about three months after admission, one month before the first contact with their children in the community. The parent groups were divided into groups for `new` parents and groups for more experienced parents. In the first groups, parents were mainly given information. In the more advanced groups, more confrontation and sharing among the parents took place.

Significantly, in Phase III, the staff decided to utilize the treatment experiences of ex-addict workers from the United States and Britain who then functioned as consultants (adjunct program advisers). Initially they instructed the staff on how to use the treatment tools of the concept for therapeutic communities, such as confrontation groups, the learning experiences and the implementation of the hierarchical resident structure.

Signs or special outfits were used as learning experiences, intended to extinguish the negative behavior while reinforcing the positive behavior at the same time (Kooyman, 1979). The learning experiences were never presented as a punishment, but instead as a natural consequence for either unacceptable or undesirable behavior. Whenever the staff decided to create a learning experience for an individual, there was an explicit description as to what was inappropriate and what could be done to improve the behavior performance. Techniques used in therapies developed in the human potential movement were added to the groups, such as bio-energetics, Gestalt and dynamic meditations.

At the end of this phase, the psychiatrist left for a two month visit to several drug- free therapeutic communities in North America, among which were Daytop Village, Phoenix House, Odyssey House, Project Return and AREBA in New York and Delancey Street, and Synanon at the West Coast and X-Kalay and Portage in Canada.

Phase IV: The post professional phase (January 14, 1975 - June 21, 1976)

At the start of this phase, most of the original staff had gone or had changed their position. The psychologist and his wife had returned to their country of origin to set up a therapeutic community. A graduate of the Emiliehoeve had become assistant director of the therapeutic community. The daily running of the program was delegated to him. More and more the Emiliehoeve staff became a mixture of professionals and para-professionals (graduates of the program).

In the Emiliehoeve the residents were divided into different groups according to their level of progressAt this point these groups, called `phases', were established. Before entering a new phase, residents were interviewed by staff members and residents of the next phase. The residents who were new, had separate 'prospect' groups. They were given an older brother or sister after they had been admitted, to introduce them into the new situation of living in a drug-free community. The community residents were divided into three groups; the prospects (who became members of the community after a month), the middle group and the older residents. Separate so-called 'status' group meetings were held within these peer groups at regular intervals, usually monthly, to discuss each member's progress. Shortly before entering the re-entry program, a resident was given re-entry candidate status. This resident had to write a report on his stay in the therapeutic community and was usually placed outside the resident structure for a short period, having no specific responsibilities within the community. The re-entry program was given a separate staff. Weekly group sessions and house meetings were run in the re-entry house.

In May 1975, the staff and the ten residents of a new therapeutic community which had started some months earlier in Utrecht were taken over by the Emiliehoeve. The residents were regarded as if they had been Emiliehoeve residents since their admission. Four of them had been already residents of the Emiliehoeve. They had been transferred to the new therapeutic community At Utrecht, at its start. The staff of this therapeutic community had been trained at the Emiliehoeve and became part of the Emiliehoeve staff. This therapeutic community had not been able to survive, due to problems with the neighborhood.

During this phase it was concluded that not all addicts needed a therapeutic community for their treatment. Especially for the younger residents a day center could be more appropriate. In January 1976, a drug-free day center, 'Het Witte Huis', was opened. The psychiatrist who was the director of the Emiliehoeve also became the director of this drug-free day center in The Hague. This center was modelled after the Emiliehoeve program. In this center, like in the staff of the Emiliehoeve, graduates of the Emiliehoeve were appointed as paid staff.

The psychiatrist became the director of the drug-free program in The Hague which was part of three different foundations:

- The Witte Huis was part of a treatment center for addiction
- The Emiliehoeve part of a general psychiatric hospital (Bloemendaal) and
- The re-entry house was part of a separate foundation (Maretak).

At the end of this phase around thirty residents were living in a farm meant for ten residents. Plans for a new building were made. Nine months after the first ideas, building plans were approved and for the first time in the world a building would be erected for the purpose of housing a therapeutic community for addicts. This building was to be built with natural materials, bricks and wood, with large open spaces in the building and large bedrooms. Although a modern building, it had a warm and friendly appearance.

Phase V: The closed community phase (June 21, 1976 - April 21, 1977)

This phase can be characterized as a closed community. From June 21, 1976 the position of the director of the therapeutic community was no longer held by the psychiatrist. A British ex-addict who was already working in the Emiliehoeve Therapeutic Community program as a consultant in Phase IV, temporarily assumed this position. This director who was not able to speak Dutch, had little or no contact with what was going on in the world outside. society. In fact the message given to the residents was: „Society is bad, our community is good; the residents have to become change-agents with the task of changing society outside after their graduation."

Discipline was extremely strict. Learning experiences such as signs and bald heads (also for women) were used more and more as punishment. Instead of sitting on cushions, the residents sat on chairs during the encounter groups. Closed weeks, centered on one theme, were held during which no new residents were admitted. Residents exercised strict control over each other. Written reports were given to the staff daily or weekly. The director could control some parts of the therapeutic community from his room by means of the monitor of a closed circuit camera. One of the slogans used as a basic philosophy was „If you are not part of the solution, you are part of the problem."

A teacher was added to the staff, who gave education lessons on Saturday morning for all residents, and classes during the week for those who needed basic education, among them two 14-year old residents. Although these young residents were given special attention, it became clear that they did not fit within the structure of the therapeutic community. The same could be said of the two occasions when young children together with their mothers were admitted to the therapeutic community, during this period.

On August 31st the new building was opened by Crown Princess Beatrix. The next day a national conference for staff as well as residents from all drug-free therapeutic communities in the Netherlands and Belgium was held on the Emiliehoeve premises. At the end of that day, two re-entry residents openly gave up smoking followed by many others. Without consulting the other directors or the program director, the director of the Emiliehoeve TC declared the Emiliehoeve tobacco-free and no-smoking a cardinal rule for the residents. Hesitating, the other directors of the other therapeutic communities still present followed this example. This no-smoking rule did not lead to an increase in the premature departure of residents from the therapeutic community. The residents just had one more thing they had to change. Besides, it was not so difficult to stop in an environment were tobacco and ashtrays were absent. Visitors were asked to leave their tobacco in bags at the entrance of the Emiliehoeve premises. However, most probably the no-smoking rule kept more potential residents out of the Emiliehoeve than otherwise had been the case. A reluctance to seek admission to the Emiliehoeve was definitely the result of the strict and inflexible discipline and extreme measures such as shaving heads. The therapeutic community got the reputation of being a concentration camp in the street drug scene. This reputation was fed by former residents who left prematurely and returned to the drug-scene, and by an article in a magazine describing extreme situations such as, residents wearing signs for weeks, the copper name-plate on the door of the director, and a ban for some residents to speak to others for more than one day.

The residents had less direct contact with the staff than before. They were expected to treat the staff with great respect. They obeyed the staff out of fear. The distance between staff and residents increased. The director of the therapeutic community had developed an almost paranoid attitude towards the outside world and towards professionals in particular. The program director finally realized that there were serious signs of abuse of power by the director and that he had to stop this process. When the director of the therapeutic community denied older residents the opportunity to confront him in an encounter group, the program director felt that he had to take immediate action. He decided not to wait until the director's contract had ended but to call the director of the therapeutic community to the re-entry house, where he discharged him immediately and also forbad him to return to the therapeutic community. What made this decision also inevitable was the plan of the Emiliehoeve director to stop sending residents to the re-entry house because he did not trust the re-entry program. That night the program director introduced a completely new staff to the Emiliehoeve residents as every old staff member had chosen to leave with their director. A graduate of the Emiliehoeve program who had been a staff member of the day center Het Witte Huis since its start, was appointed director of the therapeutic community. The social worker who had set up the parents program was appointed as co-director. The director was responsible for the treatment, the co-director for the management and organization of the program. After having explained the situation to the residents, they all went into groups with the program director and their new staff. Only four residents left during that week, two of whom returned soon after. The next phase had started.

Phase VI: The integration phase (April 21, 1977 - May 14, 1980)

In this phase the distance between staff and residents became much smaller. An integration took the place of the structure and techniques of the American self-help model and the professional input. Especially in the groups. In the groups more attention was paid to past experiences and fear of intimacy than in the previous period. The residents were seated again on cushions during groups instead of chairs and bonding therapy techniques which had not been used in the fifth phase were re-introduced.

The period during which new residents had to spend wearing overalls was reduced from 2 months to one month. The time residents were not allowed any contact with the outside world was gradually reduced from 4 months to 2 months. Sexual contact between residents was allowed after it had been discussed with the staff and their peer group and after residents had been in the program for four months. They could apply to sleep together for a certain night in a room in the staff building. Although the applications were rare as a result of this policy, sexual acting-out between residents which had happened before, disappeared. Relationships that developed between residents during this phase were rarely challenged and were more or less sanctioned by the staff. Most of these relationships turned out to be a serious handicap for treatment. After discharge, these relationships usually did not continue and both persons relapsed into drug abuse. In other cases, the couple split from the program when their relationship was challenged.

Newly admitted residents were called „eggs" and two persons, a male and a female resident, were made responsible for the „eggs" in the „egg box". During this period two staff members who had been trained in the Institute for the Training of Addiction Therapist, which had been started in September 1976, on the initiative of the Emiliehoeve, joined the Baghwan sect. They followed the example of the Institute's director, the American, who had been consultant of the program in the second and third phase. In meetings of staff from the various units of The Hague Drug-Free Program, consisting of the Induction Unit, the Emiliehoeve therapeutic community, the re-entry program Maretak, the day center Het Witte Huis and the Prevention Unit, the following policy was decided upon; staff members in the program were not allowed to talk to residents about their personal conviction, beliefs or political choices in order not to influence the residents' own opinion. It was considered to be against the program's goal, which was to improve the client's ability to make independent choices. The two staff members who had joined the Baghwan Movement were told not to wear their orange clothes and mala (chain with portrait of Baghwan) while working, not to use their Baghwan name and to accept the program policy on personal convictions. One of them agreed to this policy, the other one found this unacceptable for himself and left the program.

During this phase the staff of the different units of The Hague Drug-Free Program worked together, holding combined meetings. Survival trips in the hills of South East Belgium, organized by an Outward-Bound Foundation which had become part of the residents program in this phase, were also organized for the entire staff. The no-smoking rule was abandoned about one and a half years after its start. The main reason being that it was too difficult to handle for some staff members who had not given up smoking. Shaving heads was abandoned for women and at the end of this phase also for man. Although head shaving was the consequence for having left the program to use drugs, and may have acted a deterrent which kept some residents in the program after they had been re-admitted, this extreme measure had probably also discouraged many other addicts from applying for admission.

Since the Summer of 1978, methadone was made easily available in The Hague in so-called low threshold programs. As a result of this, it became more difficult to require clients to go to the induction center, an ambulatory unit to stop taking drugs and to come back clean the next day. Although some clients were successfully detoxified in the ambulatory methadone detoxification program, run at the out-patient center for addiction by the psychiatrist who was also the program director of the Drug-Free Program, for a growing number of addicts this was no longer a possible option. It was much harder to go through the withdrawal phase using methadone than using heroin or amphetamines without methadone. To make it possible to go through the withdrawal phase in an in-patient center where also the induction for the therapeutic community or drug-free day-center was held, plans were made for a detoxification center within the program following the example of the Heemraadssingel Center established in Rotterdam and linked with the Essenlaan TC. On May 14th 1980, a 12-bed detoxification unit, called 'De Weg' was opened within the Bloemendaal organization as part of the Drug-Free Program. All induction of Emiliehoeve residents now took place in this center, except for those coming from other detoxification centers and for candidates for admission who were in prison. With the opening of De Weg the next phase started (Wagenaar 1981). For the purpose of this research, this VIth phase ended on August 20th 1977; the day after the admission of the 250th resident admitted to the Emiliehoeve for the first time.

Phase VII: The open program phase (May 14, 1980 - February 14, 1981)

In this phase the Emiliehoeve TC was highly involved in preventive programs promoting the values of a drug-free life-style. Staff and residents participated in training programs at prison and police training institutes, academies for social sciences, and the training of medical students and persons working in other programs for addicts.

The combination of a director, responsible for the treatment program and a co-director, responsible for the management and staff, had not worked satisfactorily in the therapeutic community. In the spring of 1981 the co-director of the Emiliehoeve became the director of the re-entry program Maretak and the director of the Emiliehoeve left the program. Temporarily the program director took over their positions in the therapeutic community.

More residents from minority groups were admitted during this phase, especially from the Surinam population. When their number had reached a group of five residents, mainly seen at induction by a Surinam re-entry resident, their tendency to drop out early diminished. Since then, a minority group of Surinam residents has been present at the Emiliehoeve Therapeutic Community.

In this phase the treatment program stabilized. The progress of each resident was evaluated by staff and peers before his advancement to a following phase. A scheme was made describing the requirements necessaryto be nominated for promotion to each following phase. After the first phase of about two weeks during which the residents were now called „youngkies", four other phases of around three months to four months followed. There were 8-hour probes with peers 3 to 4 times a year, for the first-phase peer groups („the frogs") on the past, the second-phase peer groups („the kangaroos") on the here-and-now, for the third-phase peer group („the fanatics") on themselves and for the fourth-phase („the diamonds") on the future. It was compulsory for residents in the fourth phase to become members of a club outside of the therapeutic community to experience contacts with non-using persons of their own age group.

Before going to the re-entry phase the residents were part of a re-entry candidate period of around two weeks. During that period they were called „the pioneers". They were given separate sleeping quarters in the therapeutic community. In this last period the resident wrote his own evaluation of his experience in the therapeutic community. In the therapeutic community, combined group therapy sessions were held for residents of the therapeutic community in their pre-re-entry phase and residents in the first phase of the re-entry program.

In the re-entry program the resident spent roughly two weeks in the house, followed by a first phase of around three months, during which he worked as an assistent to the staff in the program, followed by a phase during which the resident worked as a volunteer outside the program. During the last phase the resident worked full-time at a job and moved out of the re-entry house to his own room or apartment. In re-entry the residents who did not have a problem with alcohol before their admission to the program, were given an alcohol-learning period of four weeks in the second phase of their stay in the re-entry program. In this period they could experiment with drinking alcohol outside the re-entry house. This period was followed by an evaluation of their peers to work out if they could or could not handle the use of alcohol.

Although sexual relationships between residents were possible in the way as described earlier, these relationships were not encouraged and were now only allowed after a resident had spent six months in the community. During this VIIth phase, it was possible to open a second re-entry house, expanding the capacity to 20 residents in the two houses. On the ground floor of the second house, an information center on drug prevention was opened, providing information on treatment programs and prevention activities. On September 1st 1981 the director of the detoxification center De Weg, a graduate of the Emiliehoeve program, became the new director of the Emiliehoeve. The program director who had been the founder of the program announced his departure to become the medical director of the multi-modality treatment organization for drug addicts and alcoholics of the Jellinek Centrum in Amsterdam. This took place on February 14th; on the 10th anniversary of the program.

Phase VIII: The adult program phase (February 14th 1982 - now)

After the program director left, this function was no longer continued in the same way. One psychiatrist took over the complete medical responsibility for the treatment programs at the detoxification center De Weg, the therapeutic community Emiliehoeve and the re-entry program Maretak. Another psychiatrist was appointed as the psychiatrist responsible for the treatment in the daycenter Het Witte Huis. As Het Witte Huis was part of a different treatment organization, this program lost its close connection with the Emiliehoeve program. The Friends of the Program Organization however, stayed linked with both programs.

The Emiliehoeve and De Weg became more integrated in the Bloemendaal Hospital Organization. The management of the therapeutic community improved. In the treatment program of the Emiliehoeve greater stress was laid on the residents' own responsibility to change. The period that new residents had to spend in overalls was first reduced to two weeks, then abandoned. From 1983 wearing an overall was only used as a special learning instrument. In 1988 wearing the large signs was also abandoned; these were replaced by small buttons or badges. Bonding psychotherapy groups and survival trips for groups of residents became an integrated part of the treatment program. From 1989 bonding therapy groups have been held weekly.

During this phase in the development of the program it was seen, more than before, that residents themselves had to make the choice to stay in treatment. The rule, for instance, that new residents should never be left alone, a vestige from the rigid closed-community phase was abolished. The transition from the detoxification center De Weg to the therapeutic community has become a more gradual process. In weekends residents of De Weg visit the Emilie-hoeve. On weekdays they can meet in sportsgames between the two groups. Residents of the Emiliehoeve run seminars in the detoxification unit. While new residents are usually admitted to the therapeutic community in pairs, they no longer have to make an emotional investment, such as screaming for help loudly in a large group of residents and staff. Instead they are seen in an interview by some staff and residents. The emotional investment is postponed to the transition into the next phase. One staff member is held responsible for guiding the new residents. He can see them individually, if he or the resident finds this necessary. New residents do not have to change their appearance at admission. Instead they may be confronted on how they look by their peers and through this, gradually change.

The residents of the Emiliehoeve program pass through four phases during their treatment in the therapeutic community, preceded by a two-week period as a junior member. After the two introductory weeks, residents are in phase one for about two months. In this phase they deal with telling their life story, making an overview of their debts and other issues of their past. The residents learn to deal with a regular day program, working in different departments. After two weeks a resident is given some pocket money. After one month, personal belongings such as make-up and jewelry kept by the staff are returned to the residents. After two months a resident can ask for privileges, such as using the telephone, and receiving visitors and mail. At the end of this phase, and of all following phases, the resident writes an evaluation and a request to proceed to the next phase. This is followed by an interview with a staff member.

The second phase is a period of about three months. The experiences within the group are given most attention. In the beginning of this phase, privileges can be requested, such as leaving the community for one day in a weekend, together with an older resident. There is a confrontation meeting with parents or other relatives with a staff member present.

In the third phase of around three months the attention is mainly on the person himself and how to deal with friendship and intimacy. In 1988 bonding groups had been reintroduced, which previously had almost disappeared. One year later bonding groups had become integrated in the program. They were held once a week. In the community the resident fulfils a job with more responsibility such as that of a department head, expeditor or coordinator. In this third phase a resident can spend a weekend outside the therapeutic community accompanied by an older resident.

In the fourth phase of three months the resident takes steps to prepare for the future. He can do some volunteer work outside the program or perform some tasks in the induction or prevention program. He can spend weekends outside alone. An evaluation of the total time spent in the Emiliehoeve is made by the resident, followed by a phase interview with the director of the Emiliehoeve, the director of the re-entry program and other re-entry candidates (Kooyman, 1992).

The parent groups were extended to the first months of their children's admission to the Emiliehoeve, as well as to parents of clients admitted to the detoxification center. At the Emiliehoeve as well as at Het Witte Huis, cases were selected for family therapy sessions. There were also special group sessions for brothers and sisters of the residents, for residents and their (drug-free) partners and for residents with children. A teacher comes once a week to prepare residents for going back to school or to provide basic education. A theatre class is held from time to time, run by volunteers.

The transition to re-entry became a more gradual process. Residents of the fourth phase attend a special meeting once a week run by the re-entry staff to discuss various topics, such as free time planning, education, relationships, parents and sexuality. In the last four weeks in the therapeutic community before going to re-entry the resident spends one afternoon in the re-entry house for orientation. He has to make the necessary arrangements for his new situation. He has to join a club outside the program. He invites his parents to meet the re-entry staff. In the first phase of the re-entry program the client participates in the re-entry encounter group, held once a week, and he also attends the parent group. After five months he may have his alcohol-learning period. After eight months he leaves the re-entry house. He can come to the re-entry encountergroup at least once every two weeks. He may have found a job or have started with a course of study. This second phase of the re-entry program lasts three months. Then follows a month in which preparations for the graduation are made.

A serious set-back for the program was the decision of the city of The Hague in 1984 to stop subsidizing the Maretak Foundation. The arrangement was that the Bloemendaal Hospital Authority would set up a re-entry house as part of their organization. The information and prevention center moved to one floor of one of the Maretak re-entry buildings. Temporarily the re-entry residents were housed inside the premises of the psychiatric hospital in former apartments for nurses, situated on the second floor of a psycho-geriatric ward. The Maretak Foundation dissolved. On February 1989 the Friends of the Program Organization felt that this temporary situation was no longer acceptable. A new separate foundation 'Steigers' was founded to raise money for a re-entry house in the city, in which re-entry residents would rent rooms and the Emiliehoeve staff would run the treatment program. Half a year later the re-entry residents were able to move into a better house on the Bloemendaal Hospital grounds. This house was situated on the main road passing the hospital. A year later, in September 1990 the residents were able to move into a re-entry house in the city of The Hague, bought by the Steigers Foundation with funds acquired from many sources, among others from the city of The Hague.

The program has been limited to a duration of two years, one year in the therapeutic community and one year in the re-entry program. In the last ten years the number of residents from minority groups has increased to sometimes one third of the population. The number of residents of the therapeutic community has stabilized at around forty residents. Around fifteen residents have already participated in the re-entry program.

Up to one third of the residents of the Emiliehoeve were now coming from minority groups, reflecting the growing drug problem of immigrants in The Netherlands. Besides Surinam residents also Moluccans, Antillians and Maroccans were admitted. They did fit in rather well in the program. Although the staff had to be aware of cultural differences it is obvious that the therapeutic community is a culture in itself, and that admission is a culture shock also for residents born and raised in Holland.

A new problem is how to work with residents, who are seropositive. Addicts who have already been diagnosed as having AIDS are not admitted, as the program cannot cope with the fact that they cannot participate in all activities. Besides, it is questionable, if AIDS patients should be in a therapeutic environment focusing on a future life that these people cannot be expected to have.

The drug-free programs of the Emiliehoeve, De Weg, the re-entry program and the information and prevention center, were organized as a separate unit within the Bloemendaal Organization. The director of the Emiliehoeve, who had followed professional training after his graduation from it, became the responsible head of management and staff; the psychiatrist is responsible for the treatment program. The directors of the therapeutic community and the re-entry program are graduates from the center with professional training, and the head of the detoxification center is psychiatric nurse. The information and prevention center became part of the responsibilities of the coordinator of the re-entry program.

The Emiliehoeve has become closely linked with the Erasmus University in Rotterdam. Its data base of more than a thousand residents is extremely valuable for future research.

Phase differences and research

The development of the Emiliehoeve program through different phases made it possible to investigate differences in the outcome of the different phases. It took ten years to develop a stable 'adult' program. Mistakes were made by the staff during the different phases such as delegating power to the residents at a stage when they were not ready to assume the responsibilities, or copying the tools and techniques used in the American therapeutic communities without taking into account that these originated from a very different culture. However, having learned from these mistakes, the staff could then clearly choose alternatives. During the first ten years new staff came and went. The psychiatrist who had founded the Emiliehoeve TC however, has been responsible for its treatment program throughout this period.

In the first phase the staff did not show direct emotional response and did not set clear limits. In the second phase there was direct emotional response without a rigid structure. In the third and fourth phase increased attention was given to discipline and structure. In the fifth phase most of the staff's attention was directed to changing the residents' behavior, while the staff remained more distant. The structure of the therapeutic community had been extremely rigid. Residents tended to act out of fear of possible consequences for negative behavior. In the sixth phase the structure was less rigid and the staff was easier to approach. In this phase emotional contact grew between staff and residents as well as mutual concern. A balance developed between concern and limit-setting to negative acting-out. In the following phases this important equilibrium has been maintained.

The Emiliehoeve and society

The Emiliehoeve had been started as an alternative for psychiatric hospitalization and methadone maintenance programs that had failed to stop the addicts from taking drugs. The Emiliehoeve program was at its onset influenced by the democratic movement that had started in the sixties. The principles of the democratic therapeutic communities formulated by Maxwell Jones, fit in with the concept of sharing and decision-making by consensus. Introducing the hierarchical residents and staff structure was against the common ideas of social workers and other professionals in the 'helping professions'. Their initial approval of the initiative to start a treatment center different from the current medical model approach, changed into heavy criticism of the cruel measures carried out in the therapeutic community. Part of this criticism was justified. The Emiliehoeve had become in the mid-seventies a society of its own, copying measures such as shaving heads from the American programs without considering the impact this would have on the image of the program in the Netherlands.

Another curious phenomenon became apparent. When, in the first years little or no positive outcome was seen as a result of the efforts to treat the residents, staff from other forms of treatment programs for addicts showed their sympathy. This sympathy changed into scepticism when later good results could be shown.

In the psychiatric hospital, the Emiliehoeve TC was a strange phenomenon. The staff of the Bloemendaal Hospital looked at the residents with apprehension. When in the first months residents were passing down the road along the houses of the hospital personnel, kids who were playing outside were called into their house out of fear for the addicts. This attitude changed after a children's party was held at the Emiliehoeve farm. All the children from the neighboring houses were invited to attend with their parents. The puppet-show was a success. The hundred ice-creams had just been enough for all children and the parents realized that the residents were not using drugs and were not dangerous. In later years the hospital management did not initially accept that ex-addicts, working as staff aides at the Emiliehoeve, would have lunch in the Bloemendaal personnel restaurant. Influences from the Emiliehoeve could be seen in Bloemendaal. The idea of having ex-patients working as staff was new and took some time to get used to. Within the Bloemendaal Psychiatric Hospital, a unit for adolescents adopted parts of the structure and techniques of the Emiliehoeve TC. In this therapeutic community the residents were given more responsibility in their program.

From 1975 paid graduates from the center could become staff members in the Emiliehoeve program. It was new that 'ex-patients' were accepted as staff and that their own experience was regarded as useful for their work. The acceptance of the Emiliehoeve program as a unit that could be regarded as a special kind of psychiatric treatment, had the advantage that the program could be financed according to the patient rates fixed for patients treated in specialized psychiatric hospitals.

A disadvantage of being part of the health care system was the government's policy to decrease the number of beds in psychiatric hospitals, including special hospitals such as addiction clinics. Because of this, the Emiliehoeve could officially not increase its number of beds to more than 45 (Kooyman, 1986d). So a limit was put on expanding the program. This was one of the reasons for setting up a day-center. It was started in 1976 within a separate foundation that was running a medical model clinic for alcoholics in The Hague and a consultation center and outpatient clinic for drug addicts and alcoholics. This day-center, Het Witte Huis, was set up by Emiliehoeve staff and was opened on January 15th 1976. The program was at the onset almost a copy of the Emiliehoeve with a therapeutic community structure held on weekdays from nine to five.

The Emiliehoeve had much influence on existing and new treatment centers in The Netherlands. Almost all therapeutic communities for addicts and alcoholics in The Netherlands were modelled after the Emiliehoeve. Some, such as the Breegweestee in the North, the Essenlaan in Rotterdam, the H.A.D. therapeutic community and Welland in the South, the Krauweelhuis in Amsterdam and the therapeutic communities for alcoholics Hoog Hullen in the North and Leefdaal in the South, had similar programs. They formed in 1975 a Federation of Drug-Free Programs, that existed for about five years, with the aim of cooperating in staff training, staff exchange, organizing workshops and improving contacts with the central government on common interests. This federation was later transformed into a more loosely organized section of the National Organization of Programs for the Care of Drug Addicts and Alcoholics, thus leaving the different therapeutic communities more space to develop their own identity. The initial goal of the Federation to have similar program structures, making it easier to exchange residents and staff who had only been temporarily employed. In later years differentiation of programs was seen as a natural development. It also had the added advantage that different programs could have different needs in relation to client populations.

The Emiliehoeve program was not only used as a model by other therapeutic communities in The Netherlands. Staff from programs in Belgium, England, Germany, Austria, Switzerland, Norway, Sweden, Indonesia, Italy and Greece were trained at the Emiliehoeve. The Emiliehoeve staff was active in organizing international meetings of therapeutic community staff. In 1980 the 5th World Conference of Therapeutic Communities was organized in Noordwijkerhout, The Netherlands. During this conference the World Federation of Therapeutic Communities was established.

The Emiliehoeve program has also influenced treatment programs other than those of therapeutic communities for addicts or alcoholics. The notion that addicts can be regarded as being responsible for their choice to continue to use their drugs became widely supported and it replaced a common view in The Netherlands of seeing addicts only as helpless victims. The Emiliehoeve was involved in setting up programs to prepare addicts for drug-free treatment such as crisis-detoxification centers and programs in prisons.

Great was the influence on the involvement of parents, partners and other relatives in the treatment program. The Emiliehoeve had been the first treatment program in The Netherlands involving parents by organizing separate parent meetings. Involving parents in the treatment program became the policy of many treatment programs.

Staff, residents and relatives of residents have been active on several occasions by showing their concern about proposals or plans by local governments or local and national politicians, directed at controlling or regulating and accepting drug abuse as an alternative to treatment and prevention programs. Several hundreds of ex-addicts from all over the country marched to the parliament buildings in The Hague in 1977 to protest against plans to dispense free heroin to addicts. There was also a protest against the distribution of methadone to addicts from buses. The plan to prescribe heroin for addicts has never been realized; the distribution of methadone from mobile units to as many addicts as possible without any demand to stop using heroin, was realized in the large cities in The Netherlands (Kooyman, 1984a). The primary goal of these programs, to reduce drug-related crime was however not reached (Grapendaal, c.s., 1991). A side effect was that these low threshold program alternatives were effective in allowing addicts to postpone their plan to seek treatment for their addiction, as is indicated by the increase of the age at admission to the Emiliehoeve since these programs were started (Kooyman, 1985c).

The above may suggest that methadone programs and drug-free treatment cannot be combined. The experience in The Hague and elsewhere has been, however, that cooperation between drug-free programs and treatment programs using methadone can be very fruitful. The program director of the Emiliehoeve had been responsible for the methadone programs in The Hague, Rotterdam and Leiden during the first years of their existence, while in the same period he was director of the drug-free program in The Hague, and later also in Rotterdam (Kooyman, 1984c, 1986). These methadone programs were both maintenance programs, using methadone as a substitute with the aim of stopping the clients from using illegal drugs, as well as being ambulatory detoxification programs. For many addicts, such as clients with serious psychiatric or somatic illness besides being addicted to drugs, methadone maintenance can be a good alternative, as they cannot be treated well in a stressful environment such as a therapeutic community. For many residents of the Emiliehoeve, methadone programs have been their first contact with any treatment for their addiction. Cooperation between drug-free programs and programs that have abandoned their ultimate goal of helping clients to stop their drug abuse, is a totally different matter and almost impossible. In The Hague the Emiliehoeve program took the initiative to establish a central intake meeting where all treatment programs were presented. At these meetings representatives from different programs informed the others on new methods for treatment and potential referrals.

Regular meetings were organized by Emiliehoeve staff with representatives of the court in The Hague, resulting in clear agreements on the admission of clients from the prisons to the therapeutic community, and rules for situations where clients who had been referred to the therapeutic community, left the program prematurely. Meetings with organizations active in drug prevention were instituted in the City Hall. The Friends of the Emiliehoeve Organization formed by interested professionals, parents and friends of residents or staff, was active in organizing meetings in and outside the Emiliehoeve therapeutic community. Their visits to the therapeutic community they were an important link with society outside the Emiliehoeve, helping to prevent the development of a splendid isolation.

Future developments

In the drug scene of the Nineties, cocaine is becoming more popular and used by more than half of the opiate addicts on a regular basis. As is shown in this follow-up study, the Emiliehoeve is definitely no less successful in dealing with addiction to stimulants such as amphetamines or cocaine than with addiction to depressants such as heroin and methadone.

A growing problem is the addiction to gambling seen in adolescents. When outpatient treatment such as family therapy or group therapy with or without partners fails, admission to a therapeutic community can be an answer. If this will be a separate program or an integration within the existing Emiliehoeve program is a question to be answered in the near future.

The development of a shortterm therapeutic community program of four or five months for poly-drug addicts who have not yet developed a junky life-style seems a desirable addition to the existing drug-free treatment programs in The Hague.

As was shown in the adolescent unit in the Bloemendaal psychiatric hospital, the structure and techniques of the Emiliehoeve can be used successfully in centers for other groups of clients, such as patients with anorexia nervosa, suicidal acting-out behavior or juvenile delinquents. Its usefulness for therapeutic communities for alcoholics was proved by the Hoog Hullen therapeutic community, a treatment center mainly treating persons with serious alcohol problems.

Since the start of the Emiliehoeve, background data have been gathered from almost all residents admitted. These data can be used in future research. They can show trends in the pattern of drug abuse in the course of time. In recent years common research projects have been developed by the Emiliehoeve staff and the Addiction Research Institute of the Erasmus University in Rotterdam. Through this research the Emiliehoeve program can keep its tradition of helping other programs to help themselves.

CHAPTER 11

The success of the program:
The Emiliehoeve evaluation study

Soon after the Emiliehoeve had been founded, a plan was formed to evaluate the treatment outcome. Background data of the residents were gathered after admission. All newly admitted residents were given an identification number. Re-admissions were given different identification numbers. Residents returning to the program after having been out of it for more than six months were regarded as re-admissions. If they had stayed out of the program for a shorter period, this was regarded as an interruption in their treatment. These residents kept the same identification number which they had received at their first admission. The factor of time spent in the program was obtained by adding up the days spent in treatment. If residents had spent less than 48 hours outside the program before they were re-admitted, these days were not deducted from the total 'Time in Program' (TIP).

To be able to gather information comparable with that from internationally used questionnaires, a questionnaire was developed, using items derived from existing American questionnaires such as used in the CODAP and DARP studies (described earlier) and in the research of De Leon in Phoenix House. The questionnaire was tested in 1974 among re-entry residents of the Emiliehoeve.

In contrast to earlier outcome research projects of therapeutic communities which lacked comparisons with control groups, two groups of residents were included in the follow-up research that did not receive any treatment in a therapeutic community (T.C.) One group consisted of persons who had stopped using drugs and alcohol during their participation in the ambulatory induction program and who had then decided to apply for admission to a therapeutic community. These were clients of the ambulatory induction center of the drugfree program in The Hague. The second group were persons who had stopped using drugs and alcohol after their admission to the detoxification center Heemraadssingel in Rotterdam, linked with the

Essenlaan TC, but who had decided to leave this center without continuing any further treatment.

Apart from these no-treatment groups, a comparison group was also chosen, consisting of a cohort of residents admitted to the Essenlaan TC shortly after this program had adopted the Emiliehoeve treatment program. The Essenlaan had been founded in 1970 as a medical model in-patient methadone clinic solely for opiate addicts. In 1974 the Essenlaan staff had radically changed their program into a hierarchically structured drug-free therapeutic community, modelled after the Emiliehoeve. The author who was the psychiatrist of the Emiliehoeve was asked to help them in this transition. He became the medical director of the drug-abuse treatment program of the Bouman Foundation to which the Essenlaan belonged and became the founder of the Heemraadssingel detoxification program in 1975. As he was also responsible for the induction program, the therapeutic community and the day center in The Hague, he could have access to the information necessary for research. Being attached to the Department of Preventive and Social Psychiatry of the University of Rotterdam, he was able to carry out the research himself with the assistance of staff members of this department and medical students who participated in the research project for five months of their study. The students were trained in interviewing persons that were included in the research. To understand the treatment in which the former therapeutic community residents had participated the students spent two weeks as residents themselves in the Emiliehoeve TC before starting their follow-up interview. The persons included in the follow-up interview were visited in the places where they were living at the time of the interview.

The follow-up study consisted of one interview each of all the persons who had been included in the original research sample for at least 6 months (180 days) but preferably for one year or more after their departure from the therapeutic community program. This means all persons who had left the therapeutic community or the re-entry program prematurely or those who had finished the treatment by graduation for at least 6 months or preferably longer (or in the case of no-treatment persons, those who had not been admitted to a therapeutic community at least six months after their intake interview).

Strict criteria for success derived from the programs goal were chosen. This goal was not only for clients to stop their addiction to drugs, alcohol, tranquillizers or sleep medication, but also for them to be able to sustain stressful situations in a healthy way and to stop any acting-out in a criminal way. Also admissions to crisis centers or psychiatric hospitals, arrests, drug-related convictions, or prison detentions were included in the criteria for success. At the time of the research, using cannabis, tranquillizers or sleeping pills incidentally and not in an addictive way, could not be regarded as deviant behavior; this type of use was regarded as acceptable and not a reason for being included in the no-success group. Also having a paid job was not included in the criteria for success as unemployment for young people was high at the time of the research.

The following criteria were chosen for successful outcome. After leaving the program or in case of the non-admittants, since the intake interview, the persons included in the study should have had:

1. No use of any hard drugs (heroin, amphetamines, cocaine, methadone, hallucinogenics, solvents).
2. If other drugs (cannabis, tranquillizers, sleeping pills) were used, the frequency of use during the last half year before the interview should have been less than once a week.
3. No alcohol abuse.
4. No arrest or prison detention.
5. No drug-related conviction.
6. No drug-related police contact without conviction.
7. No treatment for alcohol or drug problems.
8. No admission in any psychiatric hospital.

The respondents who were included in the success group had to score positive in all 8 areas. Successful treatment using the above criteria means therefore no relapse after treatment, or in the case of the non-admittants, after the intake interviews. This means that if for instance, a person had used cocaine only once, this person would not be included in the success group. To be able to look at success in a different way, the situation during the last half year before the interviews concerning the use of drugs or alcohol was also investigated in the interviews. This made the study comparable with studies measuring success by looking at a six month period before the interview, instead of the total period since treatment.

Of the Emiliehoeve sample all clients were re-interviewed that had been included in the success group. The re-interviews were to be held around three and five years after clients had left the treatment program. This would give information about the stability of the successful survivors of the first follow-up.

Seeing the persons from the sample at definite time intervals has been a problem. Many persons could only be traced after several trials by the students. A fact that made it impossible to see the respondents at fixed intervals was that the medical students who took part in the interviews were only available for the research during a period from February to July each year. Another problem was that during the nine years, different groups of students have been involved in assisting in the research project.

Although standard questionnaires were used with fixed questions and fixed multiple choice answers, having to use 38 different interviewers has been a problem. An advantage was that the students could work full-time and were given means to travel around the whole country to see the respondents in a face-to-face interview. Gathering urine samples was not considered necessary. The validity of self-reported heroin use can be regarded as high (Bale et al., 1981). Seeing the addicts in their own environment, by the interviewers who had been at the Emiliehoeve but who were not part of the treatment programs' staff, was found to give sufficient information on a possible relapse. Besides, the goal was to investigate the whole period since treatment or the first interview, and not to make an assessment of the moment.

The interviews usually took around two hours and were carried out by one student. Sometimes a second student was an observer during the interview. Some of the students participated in the research project for some time after their 5-month period as paid student assistants. They advised new students during their first few interviews. Before these interviews video-recorded training sessions were held.

At the intake interview addresses and telephone-numbers of three potential informants apart from parents or partners were gathered, who could possibly be contacted later to find out the client's whereabouts. These addresses could be of relatives and friends but also of several workers or staff from other treatment organizations. This information proved to be very helpful in tracing the respondents.

Not all the information gathered during the interviews has been used in this study. The student groups also produced some minor studies which were printed as student publications by the Department of Preventive & Social Psychiatry. The selected items of the intake interviews and of the follow-up interviews were statistically analyzed. Persons who had not been addicted to drugs or alcohol were excluded from the sample, as well as persons from the therapeutic community samples whose treatment had taken place in more than one therapeutic community; those who had been transferred to the Emiliehoeve or Essenlaan to or from another therapeutic community. Also those who had not completed a follow-up interview were omitted from the analysis.

To compare potential outcome differences between the different phases in the development of the Emiliehoeve TC, the Emiliehoeve clients were allocated to the phase of the program during which they had left the program, unless they had spent 75% or more of the total period of treatment in the program in one or more earlier phases. In that case they were allocated to the phase preceding the one during which they had left the therapeutic community.

To investigate the influence of the participation of parents in the parent groups of the Emiliehoeve program, attendance to parent group-meetings of at least one parent was examined. This applied to residents from phase three to six when parent meetings had become an integrated part of the program. Parents' attendance at parent groups was divided into non-attendance, attendance to one or two groups, and attendance to three or more groups.

During the gathering of this information the psychiatrist, who is the author of this book and researcher of this study, was the director of the treatment programs involved. This made it easier to acquire the necessary cooperation from the staff to implement this research which covered many years. During the period that the results were analyzed, the author/ researcher had left the position of program-director and was no longer responsible for the treatment program. This last fact diminishes a potential bias in the interpretation of the results (Kooyman,1992).

The Emiliehoeve population

In the subsequent years of the Emiliehoeve program, several trends could be observed in the background characteristics of the residents. On admission information was gathered from all clients, using standard questionnaires. Of the admissions in 1972 through 1974, 71 percent of the residents had only had primary education. During the years following the education level rose (Only primary education levels in 1972-1974: 71% as opposed to 1984: 34%).

The mean age at admission increased from 20.6 years in 1972-1974 to 24.4 years in 1983 (see table 4).F

Table 4: Trends at admission to Emiliehoeve; age at admission.

Year	Average age in years at admission
1972-1974	20,6
1975	21,6
1976	21,1
1977	21,1
1978-1979	22,2
1980-1981	23,1
1982-1983	24,4

During the course of the years, fewer residents came from lower social classes (estimated), the percentage dropped from 37% (in 1972-1974) to 33% in 1976 and 9% (in 1982-1983).

The referrals from court remained stable over the years (9-10%). More than 50% of the residents in all years had been in prison before admission. Also the male-female ratio has been more or less constant: 7 to 3. During the years 1972-1976 there were no residents from minority groups in the program. Since then the percentage rose to 17% in 1983.

There is a tendency to abuse more different drugs. In 1972-1974 an average of 2.87 different drugs was used. In 1976: 3.77 and in 1983: 3.97. Over the years there have been considerable changes in the use of the problem-drug which was most frequently used (see Fig. X). In 1972 the drugs preferred most were amphetamines, while in 1983 opiates (heroin) were the most popular drugs. At the end of the eighties cocaine became a popular drug of abuse, also among opiate addicts. While 23% to 25% of the residents in 1972-1976 had also been using inhalants to sniff, such as glue or trichlorethylene, this use dropped to 5%-9% in the period of 1978-1983. The use of hallucinogenics also dropped in the eighties. From 1976 methadone became more easily available through low threshold dispensaries from city health programs (see table 5). Since that same year the average duration of the addiction before admission increased from 3 years in 1976 to 7 years in 1983. A considerable change can be seen in the patterns of how drugs are used. Injecting drugs as the main method of use, dropped from 65% in 1972-1975 to 47% in 1980-1983. Smoking (inhaling) the drug rose from 6% in 1972-1975 to 32-47% in 1980-1983 (see table 6).

Figure X: Changes in problem drugs.

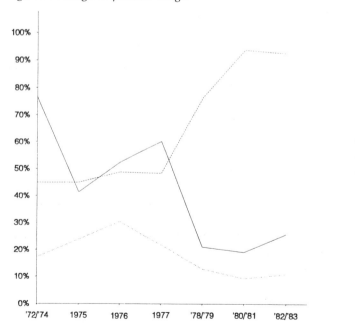

Table 5: Trends at admission to Emiliehoeve; additional use.

Year	Cannabis	Cocaine	Methadone
1972-1974	89 %	26 %	30 %
1975	80 %	63 %	50 %
1976	67 %	44 %	38 %
1977	61 %	31 %	24 %
1978-1979	67 %	40 %	48 %
1980-1981	73 %	63 %	71 %
1982-1983	75 %	72 %	75 %

Table 6: Trends at admission to Emiliehoeve; main methods of use.

Year	Injecting	Smoking	Sniffing	Swallowing
1972-1975	65 %	6 %	6 %	24 %
1976-1977	54 %	8 %	22 %	16 %
1978-1979	39 %	23 %	21 %	17 %
1980-1981	46 %	32 %	20 %	3 %
1982-1983	47 %	47 %	6 %	0 %

The sample of the follow-up study

For this research the first 250 first-time residents admitted to the Emiliehoeve Therapeutic Community since the start of the program were studied. These residents were admitted in the first five years after the program was founded. Re-admissions within six months after their departure were regarded as being part of the same admission. Re-admission after having been out of the program from more than six months were excluded from the study.

A comparison group was chosen from the Essenlaan therapeutic community in Rotterdam, consisting of a cohort of 90 residents admitted after the Essenlaan program had introduced a program modelled after the Emiliehoeve program (phase 4). This was in the period that the psychiatrist who was the director of the Emiliehoeve program had also become the director of the Essenlaan program. The Essenlaan residents of this cohort were also first admissions.

As a control (no treatment) two groups of persons were included that had made a choice to stop taking drugs and who had been admitted to either the ambulatory induction program of the Emiliehoeve in The Hague, or to the in-patient induction program of the Essenlaan in the Detoxification Center in Rotterdam. These persons had discontinued their induction program and left before admission to the Emiliehoeve or Essenlaan. The persons of this group were called 'non-admittants'.

Persons with drug problems that were not clearly addicted to hard drugs (amphetamines and/or opiates) and/or alcohol were excluded from this study. Persons who were admitted to the therapeutic community program for other problems (mainly psychiatric ones) than addiction to hard drugs or alcohol, or for psychiatric problems, were also excluded. Persons with incomplete records at admission (or at intake in case of the non-admittants) were excluded as well.

Persons transferred to the Emiliehoeve or Essenlaan from another therapeutic community, or persons who were transferred for further treatment from these therapeutic communities to other therapeutic communities, were also excluded.

Of the Emiliehoeve sample, twelve persons were excluded because the reason for their admission had not been an addiction to hard drugs or alcohol. Eleven persons were excluded because they had been transferred to the Emiliehoeve from another therapeutic community (such as the Essenlaan or the Breegweestee) or because they were transferred from the Emiliehoeve to another therapeutic community before they had finished their treatment.

Of the remaining 227 persons, 172 had been interviewed in a follow-up interview at least once after they had left the program. This is a response percentage of 75.8%. Of the Essenlaan group 67 persons could be included in the comparison group of 90 cases; the majority of the exclusions (as in the Emiliehoeve group, no addiction to hard drugs and/or alcohol, transferrals to and from other therapeutic communities and incomplete records at admission) had incomplete records. Of these 67 residents, 47 were interviewed in a follow-up interview, a response percentage of 70.1%. Of the control groups of non-admittants 49 clients were interviewed at the ambulatory induction center of the Emiliehoeve; 32 of them could be traced for a follow-up interview (65.3%). Fourteen clients were interviewed at the in-patient induction center of the Essenlaan; 12 of them could be traced for a follow-up interview (85.7%). The two non-admittants groups did not differ significantly in their response rate. Both groups were pooled, resulting in a group of non-admittants of 63 clients of whom 44 received a follow-up interview, a response rate of 69.8%.

The differences in response rates between the Emiliehoeve (EH) and Essenlaan (ES) and the non-admittants (NA) groups were not significant (see table 7).

Table 7: Response rate.

	N-total	N-follow-up	Response rate
EH	227	172	75.8%
ES	90	67	70.1%
NA	63	44	69.8%
Total	357	263	73.7%

The Emiliehoeve residents had been treated during the first six phases of the development of the program. For the follow-up study the sixth phase ended on the day after the 250 first-admitted residents entered the therapeutic community. This means that the residents of the sample who arrived last belonged to subsequent phases. Eighteen persons could not be attributed to the 6th phase, because they left after August 20th 1977, the last day of phase 6, and then also spent less than 75% of their total time in the program in phase 6. Of the total study sample (N=227), 209 clients could attributed to one of the six Emiliehoeve phases.

The response rates for the six Emiliehoeve phases are shown in table 8:

Table 8: Response rate in different EH phases.

Phases	Percentages	
1	46.2%	(6 out of 13 persons)
2	79.2%	(19 out of 24 persons)
3	73.9%	(17 out of 23 persons)
4	75.4%	(43 out of 57 persons)
5	78.2%	(43 out of 55 persons)
6	73.0%	(27 out of 37 persons)

These differences in response rate were not significant.

Also, between the two groups of non-admittants there was no significant difference in the response rate; the response rate for the out-patient group was 65.3% (32 out of 49 clients) for the in-patient group 85.7% (12 out of 14 clients).

Comparison of the background data and drug history of the three groups
Comparing demographic background and druguse histories, there were only significant differences found in the prevalence of amphetamine problems. In the EH-group significantly more subjects had problems with the use of

amphetamines than clients of the ES-group. Also the prevalence of amphetamine problems in the NA-group was higher than in the ES-group. A possible explanation is that the Essenlaan program was an in-patient methadone clinic for opiate addicts, before it was changed into a drug free therapeutic community and therefore it may have attracted less persons with amphetamine problems at the time of the study.

Clients from the Emiliehoeve group had more previous hospitalization than those in the other groups. A possible explanation might be the fact that the Emiliehoeve is part of the Bloemendaal Psychiatric Hospital organization and may therefore have received more clients who had been hospitalized in a general psychiatric hospital or psychiatric crisis center.

On all further drug history data, such as age at onset of daily use, length of daily use, alcohol problems and suicide attempts no significant differences between the two groups were found. On all demographic background data such as, sex, age at intake, ethnic group, last finished education, social class and work situation at intake, no significant differences were found. There were also no significant differences found in the criminal background between the three groups.

Differences between the Emiliehoeve and the Essenlaan populations

Differences indicating the holding power of the two programs were investigated by looking at the total time spent in the program (TIP), early drop-outs, and graduates.

Retention
The time spent in the program by EH and ES residents is measured in days. A difference between the Emiliehoeve and Essenlaan groups was that the Emiliehoeve residents in almost all cases entered through an ambulatory induction center while the Essenlaan group almost exclusively entered the program from the in-patient program at the detoxification center in Rotterdam.[1]

In Fig. XI the retention curves for all residents in the sample of the two therapeutic communities are shown.

The retention curve shows that approximately 25% of the clients of the Emiliehoeve leave treatment within a month. After about four months the number of clients that remain in treatment is around 50%. The drop-out rate

---

1. In a pilot study of a sub sample of the Emiliehoeve group this group was compared with a group of Emiliehoeve clients that were admitted after the detoxification center 'de Weg' was opened in 1980. A trend was found that the last group had less early drop-outs.

stabilizes at a low level around six months after admission. The retention curves in Figure XI are similar in shape. The ES-curve however shows a somewhat stronger decline than the EH-curve after one year.

Figure XI: Retention Emiliehoeve and Essenlaan clients.

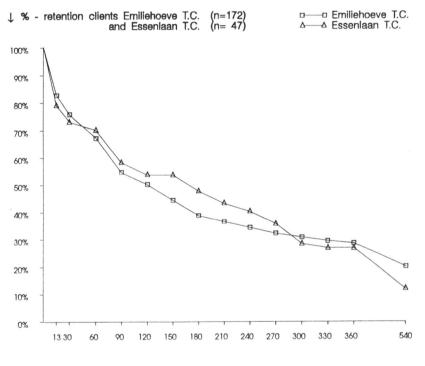

↓ % - retention clients Emiliehoeve T.C. (n=172) and Essenlaan T.C. (n= 47)

□——□ Emiliehoeve T.C.
△——△ Essenlaan T.C.

Time in program (days) →

Time in the program

The average TIP for all Emiliehoeve residents is not significantly different from all Essenlaan residents.[2]

---

2. Within the Emiliehoeve sample the question was raised about whether there was a seasonal influence on retention. The clients admitted between April 1 and June 30, and those admitted between October 1 and December 1 of the years 1977, 1978 and 1979 were compared for TIP. The difference was not significant. However, a trend could be seen that the mean TIP for persons admitted at the start of the Winter was longer than the mean TIP for persons admitted at the start of the Summer. Of the Summer group 65% had left the EH program before the 14th week; of the winter group this was 50%.

Early drop-outs

Clients leaving the program within 14 days were considered as early drop-outs. They left before day 14, day 1 being the date of admission to the therapeutic community. The percentages for the two groups are for all clients in the study (see table 9):

Table 9: Early drop-outs (<14 days) all clients Emiliehoeve and Essenlaan.

EH	16.7%
ES	20.9%
EH + ES	17.7%

There is a slight but not significant difference (chi^2=0.4; df=1; p=.55).[3]

Graduates

Of the total of 219 respondents who had been treated at the Emiliehoeve or Essenlaan who participated in a follow-up interview, 46 were graduates from the center (21%). There was no significant difference between the EH (21%) and ES (14.9%) percentages.

Clients of the first three phases of the EH program did not pass a graduation, as the re-entry program had not yet been developed and the graduation ceremony had not been established. When clients of these phases of the Emiliehoeve are omitted, the percentages of graduates of the Emiliehoeve rises to 26.6% (36 of the remaining 135 clients).

Contacts with parents

Before admission 66.5% of the EH residents had had face-to-face contact at least once a month with at least one parent. For the ES residents the percentage was 76.6%. The percentage of the NA (non admittants) group was 62.5%. The differences were not significant. Of all three groups combined 67.6% of the persons had had face to face contact at least once a month with at least one parent. These percentages refer to the last half year before the intake interview.

---

3. Hendriks found in his study that Emiliehoeve residents staying shorter than three months had a shorter employment history, less social problems, less drug problems and a higher treatment need for alcohol problems. Of persons with an anti-social personality disorder more than half had left the therapeutic community within three months compared with 28% of the persons without this disorder (based on DSM-III diagnosis). Also persons with a diagnosis of panic disorder left treatment within three months at a higher percentage (72%) than persons without this diagnosis (38%; Hendriks, 1990).

Differences within the Emiliehoeve sample

The phase of the Emiliehoeve, demographic background data and drug history
Of the first six phases in the development of the Emiliehoeve (N=209) the
background data and drug history were compared. There were no significant
differences found in: sex, age and admission, ethnic group, last-finished
education, social class or work history; in contacts with parents, and drug-
abuse related contacts with police or arrests without being convicted; nor in
admissions to general hospitals, crisis centers or other clinics related to
addiction problems; nor wheter alcohol, opiate or amphetamine addiction
was the main problem; nor in the onset of daily use, the duration of daily use
and the main mode of use.

The clients from the different phases did not differ significantly in drug
history and background date, except for the prevalence of suicide and
criminal records that were higher among the residents admitted later in time.
In the course of the years the Emiliehoeve population seems to have been
more deviant.

Comparison of the clients with follow-up and without follow-up

The subjects who could not be traced for a follow-up interview or who had
refused to take part, differ from those who were seen in a follow-up
interview only in having had a more criminal background and having more
often received previous in-patient hospital or crisis center admissions in
relation to the treatment for their addiction. So, the clients seen at follow-up
were in only a few aspects different from the clients not seen at follow-up,
regarding background data and drug history.

There was a considerable difference in time spent in the treatment
program. Clients seen at the follow up did spend more time in the program
than clients not seen at a follow-up interview. When we compare the early
drop-out rate, we see for all the clients (N=294) of the Emiliehoeve and
Essenlaan combined, an early drop-out rate of 17.7%. The early drop-out rate
of the clients who could be included in the follow-up (N=219) is lower:
15.1%. Clients who spent more time in the treatment program, apparently
could be traced more easily for a follow-up interview.

Of all the clients, 14.9% had no contact at all with at least one parent,
67.6% at least once a month, the remaining 17.4% had some contact but less
than once a month. Of the clients included in the follow-up 71.3% had
contact at least once a month with at least one parent, of those without a
follow-up interview, 58.2% had contact at least once a month with at least
one parent. The difference is not significant. However there is a trend that
persons included in the follow-up had somewhat more contact with parents
than clients who were not interviewed at a follow-up.

Time of interviews

The first resident of the Emiliehoeve sample was admitted on the date the program started (February 14th 1972); the last resident in the samplebeing the 250th admission, on August 19th 1977. The first resident of the Essenlaan cohort was admitted on March 20th 1974. All first admissions of the Essenlaan admitted since that date were included in the cohort. The last one was admitted in December 25th 1977. Non-admittants interviewed at the ambulatory induction center in The Hague were seen between November 1st 1978 and May 14th 1980 (start of the detoxification center De Weg). The non-admittants interviewed at the detoxification center Heemraadssingel in Rotterdam were seen between November 11th 1978 and September 30th 1979. As the non-admittants had been interviewed in a later period than the intake of the persons admitted to either the Emiliehoeve or Essenlaan group, the follow-up interviews could take place largely during the same years for all groups. This means after the persons had been exposed to society for at least 6 months following treatment or intake only. The follow-up interviews were only repeated for Emiliehoeve clients a second and third time, and only for those who had been successful according to the chosen criteria at the first follow-up.

Time between follow-up interview and date of leaving the program or of the intake interview in the case of the non-admittants ('Time Out of Program')

The 'Time Out of Program' (TOP) is only applicable for those subjects who were seen at a follow-up interview at least 180 days after they left either the Emiliehoeve or the Essenlaan therapeutic community. The mean TOP for the Emiliehoeve clients in the sample at the first follow-up interview is: 710 days, the median being 535 days (between the day they left the program and the date of the first follow-up). So the mean TOP was 1.9 years and the median TOP around one and a half year. The mean TOP for the Essenlaan clients in the sample is: 1090 days, the median being 958 days. So the mean TOP of the follow-up for the Essenlaan cohort was around two and a half year. The TOP of the Essenlaan clients was significantly higher than the TOP of the Emiliehoeve clients. The minimal mean TOP of the Essenlaan clients (851 days) is higher than the maximal mean TOP for Emiliehoeve clients (791 days). The time between the intake interview and the follow-up interview of the non-admittants is intermediate (mean: 855 days; the median being 804 days) and is not significantly different when compared with the TOP of either the Emiliehoeve or the Essenlaan group. The mean time between the intake interview and the first or only interview of all respondents (N=263) and the follow-up interview is 802 days, the median being 606 days. The successful Emiliehoeve clients at the first follow-up were re-interviewed at an average TOP of 2 years and nine months at the second follow-up and at an average of five years at the third follow-up.

Treatment outcome

In this chapter a description is presented of the treatment outcome results using two criteria: success since leaving treatment and success for the last half year. Comparisons are presented on the criteria of program groups and non-admittants to treatment. This analysis has been conducted on a total sample of 263 respondents who had completed a follow-up interview.In addition, the results are presented of a survival analysis based on the long-term follow-up of the EH clients who were successful at the time of the follow-up interview. Because we were interested in those clients who had no long-term relapse, only the criterion of success since leaving treatment was used in this survival analysis.

Overdoses and suicides

At the time of the third follow-up of the original sample of Emiliehoeve clients (N=227) 12 persons were reported dead. As information from the official death registration cannot be obtained, this is the minimal number of deaths within a period of five to ten years since the clients left the program. Of the 12 persons 11 were male, 7 died of an overdose of drugs, 4 had committed suicide and one died of an accident related to alcohol abuse. Of the Essenlaan sample 7 persons, 6 of them men, were reported dead.

General success percentages

Success since leaving treatment
In terms of the criterion of success since leaving treatment 25.5% (67 of 263) clients can be classified as successful. Significant differences have been found between the groups (Chi2=14.4; df=2; p<.001). For the EH-group, the highest success percentage was found; 32.0% (55 of 172). The ES-group had a success percentage of 21.3% (10 of 47). The NA-group had the lowest total success percentage of 4.5% (2 of 44). Between the two NA-subgroups no significant difference has been found (see table 10).

Table 10: General successes; all clients with follow-up.

	Total (N = 263)	EH (N = 172)	ES (N = 47)	NA (N = 44)
N Success	67	55	10	2
In percentages	25.5%	32.0%	21.3%	4.5%

Of all residents admitted for the first time to the Emiliehoeve, 32% could be classified as successful using strict criteria for successful outcome when seen at the first follow-up. Of the Essenlaan residents seen at the follow-up interview, 21.3% were successful according to the same criteria, while only 4.5% of the non-admittants could be classified as successful at the follow-up interview.

Success in last half year before interview
The criterion of success in the last half year is less strict than the criterion of success since leaving the program. No significant differences between the EH-and ES program groups exist in the success rates in the last half year. The percentages of success are 49.4% (85 of 172) for the EH-group, 42.6% (20 of 47) for the ES-group and 16.1% (10 of 62) for the NA-group (Chi2=20.95; df=2; p<.00)[4] (see table 11).

Table 11: Success in half year before interview; all clients with follow-up.

	Total (N = 263)	EH (N = 172)	ES (N = 47)	NA (N = 44)
N Success	114	85	20	10
In percentages	43.3%	49.4%	42.6%	16.1%

Of the 196 unsuccessful respondents classified by the stricter criterion of success since leaving treatment, 24% (47) did indeed have success in the last half year before the interview (see table 12).

Table 12: Success in half year before interview; of failures at follow-up.

	Total (N = 196)	EH (N = 117)	ES (N = 37)	NA (N = 60)
N Success	47	30	10	8
In percentages	24%	25.6%	27	13.3%

---

4. Note that the N of non-admittants used as a percentage base is 62 instead of 44. This discrepancy is because 18 additional cases have been included in this base that had a follow-up measurement, but had incomplete records at intake. It was decided to nonetheless to include these cases in this analysis because it provided a more conservative base for group comparisons since all of these 18 cases were failures on follow-up.

Almost half of the first admissions at the Emiliehoeve 49.4% had no relapse in hard drug use or alcohol abuse in the half year before the first follow-up interview. For the Essenlaan residents this was 42.6% and for the non-admittants 16.1%. Of those Emiliehoeve and Essenlaan residents who were classified as unsuccessful in this follow-up we see a higher success in the last half year before the interview than among the unsuccessful non-admittants (25.6% of the unsuccessful Emiliehoeve group, 27% of the Essenlaan group and 13.3% of the non-admittants group).

Co-factors of success
Successful outcome of the treatment in the Emiliehoeve and Essenlaan therapeutic communities was defined by criteria derived from the treatment program goals. A number of factors not included in the success criteria can be considered as indicators for success.

Social integration can be indicated by the level of employment or school attendance, and by social and cultural activities. Self-destructive behavior can be indicated by suicide attempts, addictive behavior by substitution of another addiction such as alcohol abuse instead of drug abuse. The following factors were investigated as potential indicators of success: employment and school situation, income, suicide attempts, substitution of addiction (between drugs and alcohol), and leisure activities.

The work and school situation had a significant relationship with the success percentage of both the Emiliehoeve and Essenlaan groups (see table 13).

Table 13: Work situation at follow-up.

	Work or school	No work or school	Total
Emiliehoeve*	100	71	171
	(58.5%)	(41.5%)	(100%)
Essenlaan	22	25	47
	(46.8%)	(53.2%)	(100%)
Non-admittants	11	33	44
	(25.0%)	(75.0%)	(100%)
Total	133	129	262
	(50.8%)	(49.2%)	(100%)

* = in one case unknown

The percentages of the Emiliehoeve and Essenlaan samples that worked or went to school at follow-up (58.5%) for EH and 46.8% for ES) were significantly higher than that of the NA group (25.0%). The 49.7% of the Emiliehoeve sample receiving social security payments is significantly lower than the 75.0% of the non-admittant sample. The Essenlaan sample (61%) lies in between.

A significant difference exists in the comparison of suicide attempts, before admission, and after treatment (see table 14).

Table 14: Suicide attempts.

	EH	ES	NA
Before admission at intake interview	41.7%	39.3%	35.0%
After admission at 1st follow-up	15.3%	10.7%	30.0%

In the Emiliehoeve sample 41.7% reported suicide attempts before treatment and 15.3% after. For the Essenlaan sample the percentages were 39.3% and 10.7%. In strong contrasts the non-admittant sample showed no significant differences in suicides before (35.0%) and after the first interview (30.0%). A significant relationship was found between success and suicide attempts. No evidence was found for an increased suicide risk as a result of the admission. On the contrary there was a significant drop in attempts after admission to the therapeutic community, compared with the non-admittants.

The percentage of alcohol problems since leaving treatment is significantly lower for the Emiliehoeve sample (28.5%) than for the Essenlaan sample (57.4%) and the non-admittant sample (54.5%). The differences between Emiliehoeve and Essenlaan clients can be explained as the result of a stronger emphasis on alcohol as a potential problem especially in the re-entry phase in the Emiliehoeve program at the time of this study. The percentage of respondents that were classified as not successful only because of alcohol problems is relatively low. For the total sample 3.8%, for the Emiliehoeve sample 1.2%, for the Essenlaan sample 10.6%, and for the non admittants 6.8% (see table 15).

Table 15: Not successful at follow-up due to alcohol.

	Total (N = 263)	EH (N = 172)	ES (N = 47)	NA (N = 44)
No succes due to alcohol	10 (3.8%)	2 (1.2%)	5 (10.6%)	3 6.8%

The results give little or no evidence of a tendency to substitute an addiction to drugs for an addiction to alcohol after treatment. A substitution of alcohol use for the use of hard drugs after treatment by the alcoholics in the sample was not seen; 19.2% of the alcoholics were classified unsuccessful for using 'soft drugs' (cannabis), tranquillizers or sleeping pills after treatment without a relapse into the problematic use of alcohol.

By investigating the leisure time activities of the Emiliehoeve sample, a significant relationship was found to exist between successful outcome and engagement in sports activities, attending youth or recreational centers, having hobbies, visiting theaters, cinemas and museums. Respondents who had been unsuccessful had been visiting centers where drugs were used significantly more than the successful ones. Participation in leisure-time activities can be seen as an important co-factor of successful treatment outcome.

Survival analysis: long-term results
The clients in the Emiliehoeve sample who had been successful according to the criteria at the first follow-up were seen in a second and a third follow-up interview, in order to investigate the stability of their condition. The average time out of the program at the last interview was 1801 days or 5 years (see table 16).

Table 16: Long term success of the Emiliehoeve sample (N=172).

	1st follow-up	2nd follow-up	3rd foll.-up
Mean time since leaving treatment	1 yrs. 11 months	2 yrs. 9 months	5 yrs.
Total number of successes	55 (32%)	42 (24%)	36 (21%)

Of the original sample a third had stayed successful on all criteria at the first follow-up, a quarter at the second follow-up and a fifth at the third follow-up.

Of the 18 persons who were no success at the third follow-up, 12 had already been no success at the second follow-up. Of the total group of 18, only two had relapsed into their former addiction to drugs; four had abused alcohol, all of whomhad previous alcohol problems before admission to the Emiliehoeve TC. Two persons had irregularly used amphetamines, one amphetamines, cocaine and hashish, and one person heroin. Four persons had been using hashish regularly for some time, one of them had occasionally used cocaine. The remaining four had been excluded for various reasons: using cocaine once, being sentenced to prison, daily use of tranquillizers or sleeping pills. One person had left the country and could not be traced for further follow-up.

During the last half year only four of the successful clients of the first follow-up were using hard drugs, one had alcohol problems, while the others had been excluded from success due to the use of cannabis, tranquillizers or sleeping pills. The persons who had no success at the second or third follow-up interview and who had been a success at the first follow-up interview, with few exceptions did not relapse completely in to their former addiction pattern and therefore can be regarded as partial successes of the treatment.

Estimation of treatment outcome of the total group of the Emiliehoeve residents in the sample of the study, including the residents without follow-up
Of the total EH-sample N=272, 172 clients were seen at follow-up, 55 were not seen. As time in program is the most important determinant for successful outcome, a prediction can be made of the outcome of those residents of the Emiliehoeve group not seen at a follow-up interview. A discriminant function analysis using a split half reliability procedure was used (Kooyman,1992). Using this statistical method the treatment outcome of the Emiliehoeve clients not seen at the follow-up was estimated based on the time they spent in the program. The results were combined with the outcome of the follow-up. The estimated treatment outcome for all clients in the sample at the first follow-up was: 74% failures and 26% successes.

Follow-up of a one-day resident population
A different way of measuring the outcome results is by choosing one particular day in a year and doing a follow-up for everyone present on that day in the therapeutic community. As short-stayers receive less treatment this method can give a good view on the cost-advantage of the treatment. In separately published research, this was done for one day at the Emiliehoeve (1-9-1975). In the first follow-up 12 of the total of 23 residents who were present in the Emiliehoeve on that day (at that time the maximal capacity as the new building had not been available and no more residents could be housed) were successful according to the strict success criteria of the Follow-up study described earlier: 11 of them had completed the treatment program (Kooyman, 1985 b).

The importance of 'Time in Program' (TIP) as the major determinant of treatment outcome
Given what is already known in the international literature of treatment outcome determinants as described in Chapter 9, it is essential, in estimating these effects, to separate the *direct effect* of the potential determinants on treatment outcome from their relationship with the time in the program, and, thus, their *indirect effect* on the outcome of the treatment. Therefore, the

research questions have been formulated as follows:

(1)   What is the relationship between a potential determinant and the treatment outcome, when controlling for the co-variation between the potential determinant and the time in program? Is it possible by means of a partial correlation analysis to estimate the *unique and direct effect* that a potential determinant has on treatment outcome, given the relationship between the time in program and the treatment outcome?

(2)   Additionally, the relationship between the potential determinant of the treatment outcome and the time in program should be measured. If, by means of simple bi-variate correlation analysis, there appears to be such a relationship, the question of causality has to be addressed: is it plausible that the determinant in question (e.g., the client's socio-economic status) affects the length of the time in program and, as a consequence, has an *indirect effect* on the treatment outcome?

The analysis was initially conducted on the first follow-up interview. A second analysis was conducted comparing the long term 'survivors' (i.e. successes) of the Emiliehoeve program with the total client population.

Time in Program (TIP) and success
Time in Program has been described as a major determinant of success in drug addiction studies, see Chapter 10. However, it is rare to find controlled studies of the TIP-Success relationship where other potential determinants are included. In this chapter, several other potential determinants of treatment outcome will be discussed respectively: parent participation in parent group meetings, Emiliehoeve phases, graduation, client characteristics, and time out of program along with the simple description of the TIP-Success relationship.

Of the EH and ES clients that had at least one follow-up 180 days or more after leaving the program combined (N=219), 37% of the variance in success can be explained by TIP (the Pearson product-moment correlation between TIP and the strictest criterion of success is $r = 0.61$ ($p<.001$). Repeating the same analysis broken down by program shows a slightly stronger relationship for the EH clients ($r = 0.63$; $p<.001$; N=172) and a rather weakened relationship for the ES clients ($r = 0.45$; $p = .001$; N=47). For the EH clients almost 40% of the variance in success is explained by TIP while for the ES clients only almost 20% is explained by TIP. Figure XII illustrates this relationship between succes and the time spent in the program.

The figure clearly shows that the longer a EH client is in the program, the greater is the likelihood that the client will be treated successfully. This relationship is linear: those clients who stay in the program for less than or equal to 30 days (N=32) have no success; the success rate increases to approximately 10% if the client remains in the program from 30 to 180 days (N=62); triples to circa 30% if the clients stays from 180 to 360 days (N=17) and increases considerably once again to approximately 70% if the client remains for more than 360 days (N=61).

Figure XII Success Emiliehoeve clients and TIP (n=172).

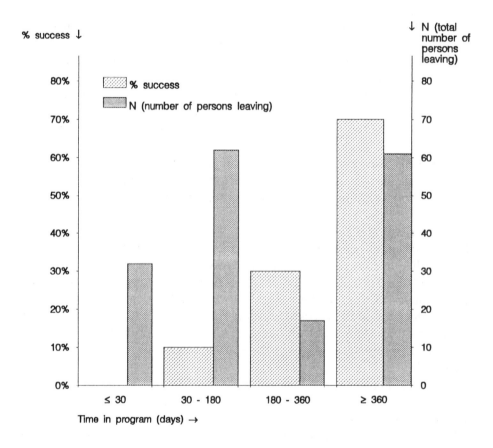

To obtain a successful outcome, a client has to remain a considerable time in the program. Leaving within a month leads almost certainly to a relapse. From then on one can see an increase in successful outcome the longer the client stays in the program. This time however is limited by the planned length of the program itself. After between 12 and 18 months the resident is expected to leave the therapeutic community to proceed to the re-entry program. The residents of the Emiliehoeve as well as the Essenlaan program then stay in a separate re-entry house for between 6 and 18 months followed by an outpatient program of 3 to 6 months during which the clients are living in their own apartment. The planned length of stay in the program of the Emiliehoeve has varied in the course of the years. The optimal time in the program is nowadays seen as 12 to 15 months in the therapeutic community and 9 to 12 months in the re-entry program. For the Emiliehoeve clients almost 40% of the variance in success can be explained by the total time spent in the program. 70% Of the Emiliehoeve clients that stayed more than 360 days in the program were successes at the first follow-up.

Success, Time in Program and Time out of Program (TOP)
From the point of view of spurious associations, the factor 'Time Out of Program' (TOP) needs to be carefully controlled in any evaluation of success, in so far as the TOP increases the likely exposure to risk factors that are not present in the protective environment of a treatment program. Given the differences that exist between EH and ES clients in TOP (longer for ES clients) and TIP (longer for EH clients), the respective differences in success between the EH (32.0%) and ES (21.3%) have been tested, controlling for both TOP and TIP. The Chi2 analysis proved to be not significant. Program success differences could not be attributed to either TIP or TOP. To further explore this negative finding, successive analyses of co-variance have been performed using TIP and TOP as co-variants. These analyses yielded similar negative results. Even when controlling for TIP and TOP, the differences between ES success and EH success still remain non significant.

Graduation and success
TIP presents a clear quantitative indicator of the exposure of a client to a given program. Another variable, graduation from the program itself, presents an indicator of a more qualitative relationship; i.e. the symbolic expression of completing the end of the program. Graduation in most therapeutic communities is imbued with a strong symbolic and ritualistic meaning. It is the planned goal of leaving the program at the optimal time. Leaving earlier means a failure to complete the total program.

For all of the EH and ES clients together with a follow-up measurement (n=219), not going through the graduation is significantly and rather strongly related to not being successful, by the strictest criterion (r=0.51; p < .001). Of the variance in success scores, 26% is explained by the graduation variable. Of the EH and ES clients a total of 46 (21%) have graduated. Of these graduates 74.5% (35) are successful while for the non-graduates only 17.4% (30) can be considered successful. The differences in these percentages is significant (Chi2 =54.8; df=1; p<.001). For the EH sample alone, 80% (29 of 36) were successful. In addition, there was no significant difference between residents leaving before graduation who had stayed longer than one year in the program and residents who graduated. Thus, there seems to be a functional equivalent of a threshold value of TIP (residency for longer than one year) and the qualitative act of graduation. However, since the above analysis does not assume any temporal sequence between graduation and TIP, an obvious question is whether graduation has any effect on the treatment outcome, controlling for TIP. Given the strong relationship between graduation and the time in the program ($r=0.869$; $p<.001$; $n=172$) it therefore comes as no surprise that, when controlling for TIP, graduation does not have a significant additional effect on the treatment outcome. Taking the relationship between graduation and the time spent in the program into account, graduation does not seem to have a specific effect on the treatment outcome. The simple quantity of exposure to the program can be inferred to be the mechanism of success. The better results of the graduates are only because they tend to be more exposed to the program.

Factors influencing drop-out and success

The high drop-out rate of the therapeutic communities is a problem. On the other hand, those who stay the longest period are the ones that benefit the most. One question still to be answered is if there is a psychological difference between drop-outs and persons who complete the program. A relationship between greater psychopathology and higher drop-out rates was found in some studies (see Chapter 10).[5]

---

5. In a cohort of the Emiliehoeve sample the residents were given a self-administered questionnaire measuring neuroticism. This so-called 'Delftse Vragenlijst' was tested on students; their mean score was 15 (Appels, 1975). The residents were asked to fill this in once every three months. There was no significant difference found between the score of the drop-outs and the residents who left the therapeutic community for the re-entry program. The score was not expected to change as it measured neuroticism. However, the score dropped from a 25,6 mean score of all residents at intake to 21.6 on the last questionnaire filled in before leaving the therapeutic community (prematurely or to re-entry). This difference was measured for those who had stayed at least 6 months in the program. This trend downwards can be partly explained by regression to normality. Both mean values are however much higher than those of normal controls.

Included in the follow-up questionnaires were open questions on reasons for leaving the community and criticism on the treatment program. Answers were compared between Emiliehoeve and Essenlaan clients. There was little difference between the Emiliehoeve and Essenlaan groups. The most frequent reasons for leaving prematurely were: injustice of the staff, rigid rules, too much pressure, lack of contact with relatives or friends, feeling alone and hard encounter groups. Graduates had more critical remarks for the staff than drop-outs. They had more criticism on the lack of free-time and most of their criticisms were on the re-entry phase of the program. Drop-outs had less differentiated criticisms on staff (mainly accusations of staff injustice).

Six hypotheses were tested in the follow-up study. The first three were:

1. The longer a resident participates in the therapeutic community, the better will be the outcome of the treatment.
2. Parent involvement helps to keep the residents in treatment and because of this improves the outcome results.
3. Lower previous education leads to a higher tendency to leave the program prematurely.

Ad 1. The longer the resident stays in treatment, the better the outcome.

This hypothesis could be verified both for the Emiliehoeve residents and for the Essenlaan comparison group. The time spent in the program was found to be the main determinant for successful outcome. In the second and third follow-up of successful Emiliehoeve clients, the amount of time spent in the program proved to have a highly significant effect on 'survival'.

All Emiliehoeve clients staying in the program less than 30 days were not successful. The success-rate increases to 10% if the client remains in treatment from 30 to 180 days, to 30% if the client stays from 180 to 360 days and to 70% if the client remains in treatment for more than 360 days. These results are consistent with the findings in literature (De Leon 1984; Holland, 1983 a; Berglund et al., 1991). Better outcome results among those clients who stay longer in the program is not only found in therapeutic communities. In a follow-up study comparing methadone maintenance clients with drug-free therapeutic community clients, Barr (1986) found that staying in treatment was in both groups related to better outcomes on drug use, criminal justice involvement and employment. As to alcohol abuse there was a significant difference: staying in treatment was associated with better outcome on alcohol abuse for therapeutic community clients but not for methadone maintenance clients. And, of course, the methadone maintenance clients were still addicted to methadone (Barr, 1986).

No optimal time in program was found. This is most probably caused by the fact that staying in a therapeutic community has a definite end. On completion of the program a ceremony symbolizing the graduates' accomplishments and departure is held. Graduates do better than persons leaving prematurely. The possibility that residents stay longer in the program than necessary has no negative outcome effect. Of the 33 graduates in the Emiliehoeve sample 28 were successful at the first follow-up (85%). On the other hand, the total duration of the program which is (including the re-entry phase) around three years, may be unnecessarily long for some residents. Short-term therapeutic communities with a 4 to 6 month program are being developed. Future research may make it possible to predict the optimal length of stay in a therapeutic community for different types of clients.

Ad 2. Parent involvement helps the resident to stay in treatment and because of this improves the outcome results.

This hypothesis could be verified. Parent groups had been introduced in the second phase of the development of the Emiliehoeve program. Parent participation (PP) was only possible in the program after the client had been in treatment for a minimum of 90 days, as in that time of the development of the Emiliehoeve program, parents were only invited to attend parent groups after the third month. Attendance of at least one parent group meeting by at least one parent was chosen as the criterion. There was almost no difference in the results when atendance of one or two times was compared with attendance of three times or more.

Fifty-seven percent of the clients in the Emiliehoeve program from phase 2 or higher, who stayed at least three months in the program, had a parent participating in a parent group. Of these clients the mean time spent in program (TIP) was 698 days compared with only 343 for those with no parents participating (see table: 17). Of those clients who had a follow-up after leaving the program (n=103) 59 (57.3%) had their parents participate in a parents group. Relating PP to success, a higher percentage of the clients who had PP showed success (64.4%; 38/59) than those who had no PP (25%; 11/44). This difference was statistically significant (Chi2=14.2; df=1; p<.001). PP has been related to TIP. Of the total (n=103) EH clients the mean TIP was 546.7 days and the median was 458 days. Broken down by PP or no-PP shows a significant difference in TIP (F=33.3; df=1; p<.001). The PP clients (n=59) had a mean TIP of 698.3 days and a median of 883.0 days while the no-PP clients (n=44) had far lower TIP mean days of 343.5 and median days of 224.5.

Table 17: Parent participation and success of Emiliehoeve clients who had TIP ≥ 90 days and follow-up (n=103).

	Clients with parent participation (N=59)	Clients without parent participation (N = 44)
Mean TIP in days	698	343
N success	38	11
Percentage of success	64.4%	25%

Table 17 displays the results of cross-tabulating the PP variable by the strictest success criterion. Of those clients who had success, 64.4% also had PP while only 25% had success without PP. The differences in the table were statistically significant (X2=7.8; df=1; p <.001).

As clients without parents or clients who had little or no contact with their parents before admission might therefore have no parent participation in the program, clients who had contact with a parent at least once a month were considered separately. For EH clients who were in phases 2 or higher (i.e. phases characterized by a parent group) and had a minimum of one monthly contact with parents a separate analysis has been conducted related to PP. PP was observed in 62.7% (47/75) of these clients. A statistically significant difference was found in this selected group in TIP on the PP variable (F=23.5; df=1; p<.001). For those clients with PP the mean TIP was 660.8 days compared to only 301.1 days for those without PP. These results are similar to those of all EH clients from phase 2 or higher who stayed for a minimum of 90 days in the program. In the following analyses the results of the parent participation of all clients are used.

The successful outcome of clients of the Emiliehoeve sample with parents participating in parent groups at least one time, was more than twice as high compared with residents with parents not participating in parent groups (64.4% as opposed to 25%). Further analysis of the relationship with the time in program revealed that this is an indirect effect (Kooyman,1992). Clients with parents coming to the parent group were staying longer in the program and were therefore more successful (see fig.XIII).

Figure XIII Simplified causal model of parent participation and TIP and success (N=103).

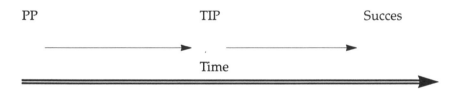

This finding is of clinical relevance; it supports the opinion that parent involvement can improve the treatment results. Whether these improved results are caused by the impact of the parent groups is still an open question. By participating at least once, a parent can symbolically give his consent to the child to choose the treatment program, while not attending may provide the opposite message to the client. The finding that parent participation in parent groups is correlated with longer TIP, combined with the experience of the low drop-out rates of the Italian therapeutic communities, where parents are already involved before admission of the client to the therapeutic community, suggests that it is worthwhile to involve parents as early as possible (Kooyman, 1987). Another reason for prolonged TIP when parents have participated in parent groups could be that the parents in these groups always receive the message to send their son or daughter back to the therapeutic community immediately in the case of the child coming home after having run away. Lastly, it can be argued, that clients who have parents interested in the therapy program have better chances because they may have a better relationship with their parents.

The strong indirect influence on treatment outcome showed that the Emiliehoeve sample in this study supports the clinical policy decision to include parent groups in the treatment program. In other research a decrease in early drop-outs was found after parents had been involved in the treatment program (De Leon & Jainchill, 1986, Nabitz & Hermanides, 1986). This suggests a program effect. The Emiliehoeve follow-up study has demonstrated, that there is a correlation between the longer time in program, when a parent participated in parent groups, and successful outcome, and that the greater success when parents participate in parent groups is an indirect effect. Residents with parents participating stay longer in the program and are therefore more sucesful. A follow-up study of the Therapeutic Community Cascina Verde in Milan found strikingly similar results on the relationship of parent participation, time in program and succesful outcome (Gori,1992).

Ad 3. Clients with a lower previous education tend to leave the program earlier than clients with a higher education.

This hypothesis could be verified (there is a significant zero-order correlation with TIP). The level of the last-finished education was measured at intake. In the therapeutic community the same tendency to drop-out of school may cause the earlier departure. It is possible that lower education is related to a lower capacity to learn from the program. It may also mean, that the person has less possibilities for work after leaving the program.

Clients of the Emiliehoeve with lower levels of education did have slightly poorer outcome results. However, when controlled for their time in program this difference disappears (see table:18). This means that poorer results of clients with lower education are mainly due to a lower TIP. However, the possibilities of having a higher level of education increase with age. When we look at the age at intake and the correlation with the length of stay in the program we see that there is clear significant zero-order correlation with TIP. The older the resident at intake, the longer the TIP. Thus the lower T.I.P for clients with lower education may be caused by the fact that they are also younger. When controlled for TIP, the age-effect on outcome results disappears completely. This means that clients with lower age at admission and lower education do have the same outcome as the others, provided that they stay. A reason for premature departure of younger persons may be that the life of an addict may be more attractive for a person who is younger. The drug-scene still being attractive may also be the reason for clients with a shorter length of daily use to leave earlier. Also here the difference disappears in outcome results after controlling for TIP. Thus although younger clients with a shorter addiction career and lower education leave earlier, they can benefit as well if they stay long enough. On the other hand older clients and clients with a longer addiction career and higher education tend to stay longer in the treatment program and therefore have better outcome results.

Another client characteristic showing a zero-order correlation with TIP is having had a prison detention before admission or intake interview. Clients with this history stay longer. Older clients have a greater chance of having been detained. This difference also disappears after controlling for TIP. All other client characteristics have no influence on TIP or the outcome results. These include: employment situation at intake, social class, contacts with parents and homosexuality. Previous convictions neither effect TIP nor treatment outcome after controlling for TIP. Female residents show a trend toward a slightly better outcome result than male residents. However, this result is not statistically significant (see table 18).

Table 18: The effects of client characteristics on the treatment outcome, controlled for TIP.

	Zero-order correlation with TIP	Partial-correlation (contr. for TIP with treatment outcome)	Valid cases
Type of drug:			
– Alcohol	0.173*	0.041	n = 172
– Opiates	0.018	-0.156*	n = 172
– Amphetamines	-0.128*	0.102	n = 172
Addiction career:			
– Age at onset of daily use	0.312***	-0.014	n = 151
– Duration of daily use (approximately)	0.224**	0.080	n = 150
– Previous treatment	-0.030	-0.083	n = 159
Criminal history:			
– Convictions	0.024	-0.088	n = 156
– Detention	0.134*	-0.004	n = 162
Demographic characteristics:			
– Sex	-0.016	-0.116	n = 172
– Last finished education	0.144*	-0.011	n = 167
– Employment situation	0.115	0.013	n = 164
– Social class	0.071	-0.079	n = 162
– Contact with parents	0.022	0.092	n = 153
– (Homo-)sexuality	-0.061	-0.027	n = 161

* - p<.05; ** - p<.01; *** - p<.001.

There are differences in the effect of the main drug of addiction on treatment outcome. Alcoholics stay longer and amphetamine addicts stay shorter in the program. Controlled for time in program these differences disappear. Opiate addicts are the only sub-group with a less successful outcome after controlling for TIP. This means that amphetamine and alcohol addicts do seem to have slightly better outcome results than opiate addicts. The significant effect of opiate use on success after controlling for TIP, however, disappeared in the third follow-up. In this long-term follow-up the initial significant difference in outcome between persons with opiates or amphetamines as the main problem drug also disappeared.

Previous treatment in clinics, general hospitals or crisis centers for addiction problems do not have an effect on TIP. There is, however, a slight trend towards poorer outcome after controlling for TIP. In the third follow-up a significant difference emerged. Clients with more previous admissions to these medical model treatment centers had less successful outcomes.

None of the variables included under demographic characteristics has had an effect on the treatment outcome, after controlling for the time in program (see table:18). However, a clear trend in sex differences and success could be observed when TIP was controlled (partial $r=-0.116$; $p=0.065$). Females tended to have more success than males. This seems to be related to the tendency among males towards earlier drop-out. The percentage differences based on a large sub-sample ($n=141$) of EH clients (Kooyman, 1985c) were: 33% of the males successful and 39% of the females. Among residents staying more than one year there were no sex differences.

Program phases and success

To study program effects on treatment outcome, the results were compared of residents treated in different phases of the development of the Emiliehoeve program. The differences between the phases are described in Chapter 10. Important differences are the introduction of encountergroups and a drug-free philosophy in phase 2 after an initial less-structured democratic phase. In the second phase there was little distance between staff and residents and strong emotional involvement of the staff. In the third phase the hierarchical structure of the American therapeutic communities was introduced and a re-entry phase was added, lengthening the program. In phase 5 the program had become highly structured and rigid. The staff was more distant from the residents and there was less attention for intimacy problems. In phase 6 a clear structure was combined with possibilities for the residents to overcome their fear of intimacy. The distance between staff and residents was reduced.

The following hypotheses were examined:

4.   Phases in the program offering a clear structure have higher successful outcome results.
5.   Phases in the program offering therapeutic learning possibilities to deal with the fear of intimacy have higher successful outcome results.
6.   Phases in the program offering both a clear structure as well as therapeutic learning possibilities to deal with the fear of intimacy have the best possible outcome results.

Ad 4. The results only mildly support this hypothesis. Phase one, with the most unclear structure, had the lowest success percentage (16.8%), one of the six clients that were interviewed had been found successful (most probably helped to stay drug-free by his membership of a religious group after treatment). However, phase five, during which the structure was very clear and rigid, was hardly more successful (18.6%, 8 of 43). The main contrasts are between phase six (51.9%, 14 of 27), with clear structure, groups on intimacy and less distant staff, and phase four (25.6%, 11 of 43) and phase five (18.6%), both having a rigid structure, especially phase five.

Ad 5. This hypothesis can find support in the results. The phases with both an emphasis on groups dealing with the fear of intimacy and also closer contact with staff members, were phase two and six. They had by far the best outcome results: 50% (8 of 19) and 51.9% (14 of 27). However, how much the phase results are due to the differences in TIP remains an open question. Especially phase six shows a strong effect of the long TIP. However, controlled for TIP, a significant direct effect was found on the treatment outcome in phase two.[6]

Ad 6. This hypothesis finds support in the fact that the phase combining a clear structure with possibilities to learn to overcome fears of intimacy (phase six), had the best results. This favorable result is related to a longer TIP. Phase two, as well as phase one, did not have a separate re-entry program so the intended TIP was shorter (in phase two, maximum 18 months).

An interesting finding in the third follow-up was that the significant effect on treatment outcome of phase two when controlled for TIP disappeared. Thus, in the long term follow-up, the initial significant difference in outcome was

---

6. In both phase two and phase six bonding therapy groups were frequently used. These groups may have a positive effect on the holding power of the community. In the therapeutic community AREBA where bonding therapy has been used since its foundation by Casriel, an extremely high retention capacity was found of 85% (Yohai & Winick, 1986).

no longer seen. A possible explanation is that the clients treated in phase two had no separate re-entry program to prepare them for the new situation in society. This lack of re-socialization possibilities may reveal their negative effects later in time.

Conclusions on the possible program effects of the different phases on treatment outcome have to be drawn with some reservations. The resident background data showed some differences. These were statistically significant on: convictions for crimes (phase 2: 13.0%, phase 3: 14.3%, phase 4: 60.0%, phase 5: 49.0% and phase 6: 47.2%), prison detentions (phase 2: 4.2%, phase 6: 47.2%) and in the number of suicide attempts (phase 2: 29.2% phase 3: 22.2% and phase 6: 64.2%). For all other client background data including educational level and the history of drug-use, there were no significant differences. Staff changes may have influenced the outcome results. The staff of the Emiliehoeve changed almost entirely several times in the different phases of the program. The person who was responsible for total treatment program, however, was the same. He also had this responsibility for the Essenlaan therapeutic community during the treatment of the persons of the Essenlaan comparison sample.

Differences in outcome between Emiliehoeve and Essenlaan

The differences in success found between the Emiliehoeve (32% of all first admissions) and the Essenlaan sample (25.5%) can be explained by the following three factors.

The Essenlaan had more opiate addicts in the sample, a shorter mean TIP, and a longer time between departure from the program and the follow-up interview. Besides this, the staff at that time of the research was less experienced than the Emiliehoeve staff. The sample of persons that did not choose admission to a therapeutic community after detoxification was clearly less successful than the persons treated at the Emiliehoeve or the Essenlaan. The outcome results are favourable when compared with follow-up results of other therapeutic communities for addicts. Outcome results in literature based on drug-use, criminal activity and employment measures range from 20% among drop-outs to 85% among graduates (Holland, 1983a; De Leon, 1984).

Different drugs and outcome

Although there is a small tendency that problematic opiate-users are treated a little less successfully, and though there is a small tendency for problematic alcohol-users and non-problematic amphetamine-users to stay in the program a little longer, it seems that whatever type of drug the EH client had problems with upon entering the therapeutic community, this does not have a substantial direct or indirect effect on the treatment outcome.

CHAPTER 12

Concluding remarks

Drug-free treatment of addicts in a structured therapeutic community with parent involvement and awareness amongst the staff of the problems of intimacy, can be regarded as a well established and succesful approach. The resident in a therapeutic community must be a participant in the therapeutic system of the community for a period that is long enough to benefit from his treatment. The longer the resident stays in treatment the better will be the result. The optimal length of stay in the Emiliehoeve therapeutic community is regarded to be at least twelve months, followed by at least another twelve months in the re-entry program. The environment in the therapeutic community is structured as a mini-society where the residents learn to assume responsibilities and to respond in a positive way to stress. This is not regarded as an environment which disables the resident from functioning in later society outside. On the contrary, in the therapeutic community residents are being prepared to function in society without the need for drugs or other substances; in the re-entry program they learn to explore and handle problems which will face them after discharge from the therapeutic community.

The various changes and differences in the treatment programs are of such a complexity, that it is difficult to draw conclusions. The results of this study indicate that less distance between staff and residents, and that possibilities for the residents to overcome their fear of intimacy, may have a stronger influence on positive outcome than a clear structure of the program. Involvement of parents is found to be strongly related to a longer period spent in the program and due to this, to better outcome results.

Treatment cannot in a causal way be given the credit for all post-treatment improvement. Influence from the family and other social influences outside the treatment setting are significant contributors. The family system of the resident in a therapeutic community was often the source of problems leading to the drugs or alcohol abuse. This was one of the reasons to prohibit any contact with relatives in the initial part of the treatment program after the induction of the resident into his new `family`, the therapeutic community. Support by the family for the choice to enter a therapeutic community and knowledge and understanding of the treatment are now regarded as important factors to keep residents in treatment. Parents who have become involved in the program will be able to send their son or daughter back to the therapeutic community if the resident contacts them after leaving treatment prematurely. The Emiliehoeve follow-up study shows evidence, that parent involvement increases succesful outcome of therapeutic community programs due to prolonging the length of stay in the treatment of their children.

Ability to deal with intimacy and sustain stable relationships can been seen as one of the main characteristics of addicts (see Chapter 3). This characteristic may lead to early dropping-out of school. As was found in previous research (see Chapter 9) it is not likely that socio-demographic variables will affect treatment. However, there may be specific psychosocial characteristics related to dropping-out prematurely from the treatment program. A characteristic may be that a dropping-out tendency, is the result of an inability to deal with intimate relationships. A career of dropping-out is then formed, leading to dropping-out of school and eventually of conventional society.

The accumulated data of the Emiliehoeve therapeutic community since the start of the program in 1972, offer great opportunities for future research. Trends in changes in drug use over more than twenty years can be monitored and program effects can be measured. Little is known of the prognosis of re-admissions which were excluded in the follow-up study. This significant issue needs further investigation. The follow-up study did show that a therapeutic community for addicts is a successful treatment approach for many addicts, not for all. Therapeutic communities cannot be a solution to the drug problem in society. No treatment can. Even when treatment is not geared to solving the addiction of the clients as, for instance, in low-threshold methadone maintenance programs, the drug problem in society will not be solved. When the goal of the treatment program is a drug-free life such as in therapeutic communities, the addict is faced with a dilemma; he has to choose between the unpleasant consequence of continuing to use and the unpleasant consequences of stopping. The choice to stop is usually made due to outside pressure and external influences, a 'negative motivation'. When the addicts' motivation comes from within himself, a personal choice of a positive drug-free lifestyle, the addict has usually already been in treatment for some months. A choice of treatment instead of prison is an example of a negative motivation. In the period of the Emiliehoeve follow-up investigation, described in Chapter 11, few clients were referred from prison. In The United States, residents of therapeutic communities sent to the program by the courts, had similar or even slightly better success rates than voluntarily admitted residents (Wexler, 1986). However, the differences found are small and legal referral does not seem to be a significant predictor to successful outcome (De Leon, 1987). In any case the relationship between specific external pressure, program processes such as parent groups and bonding group-therapy, time in program and treatment outcome, needs a considerable amount of future research. The Emiliehoeve follow-up study was partly aimed at providing a frame-work for future program and policy evaluation research.

Not all types of addicts can be expected to benefit from treatment in therapeutic communities. As the setting is the treatment, the therapeutic community offers the same type of treatment to all residents. That particular setting cannot be a suitable answer for all types of addicts. Cases requiring more individual care can be better treated elsewhere. Absolute contra-indications for therapeutic community treatment are psychoses and serious mental defects or brain-damage. A relative contra-indication is severe socio-pathic behavior. Persons that relapsed after completion of their treatment may better be referred to other treatment. A special type of therapeutic communities has been founded to treat sub-groups such as addicts with borderline personality disorders, minority groups and adolescents.Short-term therapeutic communities for addicts that have not yet developed a 'junkie' life-style and have positive contacts in society, have been developed (Kooyman, 1987). Further research is needed to be able to specify the poten-tial and limitations of different therapeutic community models.

The Synanon/Daytop/Phoenix House therapeutic community model has spread from The United States to Europe, Canada and South America, South East Asia and Australia. Although the program had to be adapted to the dif-ferent cultures, they have more similarities than differences. The therapeutic community is a culture of its own, also in the United States, the country where this model originated (O'Brien, 1981).

The Emiliehoeve therapeutic community had a great impact on the treatment of addiction in and outside The Netherlands. The folow-up study described in this book indicates the positive effects of parents involvement and therapeutic learning possibilities to deal with the fear of intimacy. Further research is necessary to investigate the potential of therapeutic communities to improve their holding power and thus to increase successful outcome.

Being successful with persons that were often traumatized in their early childhood, the Therapeutic Communities for Addicts can be a model for other client populations, especially for those that end up in institutions for juvenile delinquents and prisons for reasons other than drug-related crimes. The evolution of therapeutic communities is a dynamic phenomenon with increasing differentiation and specialization, promising new treatment possibilities for addiction problems in the future.

References

Adams, J.W. (1978). Psychoanalysis of Drug Dependence: The Understanding and Treatment of a Particular Form of Pathological Narcissism. Grune & Stratton, New York.

Alexander, B.K. (1987). The reality of drug use in the 1980's: Models of addiction and the war on drugs. In: Kaplan & Kooyman (eds.) Proceedings of the 15th ICAA International Institute on the Prevention and Treatment of Drug Dependence, Amsterdam/Noordwijkerhout, pp. 8-15. Institute forä Preventive & Social Psychiatry, Erasmus University, Rotterdam.

Alexander, B.K. (1990). Peaceful measures: Canada's Way Out of the "War on Drugs". University of Toronto Press, Toronto.

Alexander, B.K. and G.S. Dibb (1975). Opiate addicts and their parents. Family Process, 14, pp. 499-514.

Alexander, B.K. & Hadaway, P.F. (1982). Opiate addiction: the case for and adaptive orientation. Psychological Bulletin, 92, pp. 367-381.

American Psychiatric Association. Committee on Nomenclature and Statistics Diagnostic and Statistical Manual of Mental Disorders, (1980). D.S.M. III, third edition. American Psychiatric Association, Washington, D.C.

American Psychiatric Association. Committee on Nomenclature and Statistics, Diagnostic and Statistical Manual of Mental Disorders, (1987) D.S.M. III, third revised edition. American Psychiatric Association, Washington, D.C.

Appels, A. (1975). Screenen als methode voor preventie in de geestelijke gezondheidszorg. Proefschrift. Swets & Zeitlinger, Amsterdam.

Aron, W.S. (1975). Family background and personal trauma among drugaddicts in the USA: implications for treatment. Brit. J. of Addiction, 70, pp. 295-305.

Aron, W.S. and D.W. Daily (1974) Short- and long-term therapeutic communities: A follow-up and cost effectiveness comparison. The Int. J. of the Addictions 9(5), pp. 619-636.

Aron, W.S. and D.W. Daily (1976). Graduates and splitters from therapeutic community drug treatment program: An comparison. The Int. J. of the Addiction, 1, pp. 1-18.

Ausloos, G., Deslisle, C., Ecuyer, S., Lanini, Y., Rey, F. and P. Rey (1986). 50 Families d'heroinomanes. Lettres Levantines, 1. Association du Levant. Lausanne.

Ausubel, D.P. (1972). Why compulsary closed-ward treatment of narcotic addicts? Reprints. Narc. Add. Control Commission, 1, nr.5.

Bale, R.N. (1979). The validity and rehability of self-reported data from heroin addicts. Mailed questionnaires compaired to face-to-face interviews. Int. J. Addictions, 14, pp. 993-1000.

Bale, R.N., Van Stone, W.W., Kuldau, J.M., Engelsing, T.M.J., Elashoff, R.M. and V.P. Zarcone (1980). Therapeutic communities vs. methadone maintenance. A prospective controlled study of narcotic addiction treatment: Design and one-year follow-up. Arch. Gen. Psychiatry, 37, pp. 179-191.

Bale, R.N., Van Stone, W.W., Engelsing, T.M., Zarcone, V.P. and J.M. Kuldau (1981). The validity of self-reported heroin use. Int. J. Addictions, 16(8), pp. 1387-1398.

Bale, R.N., Zarcone, V.P., Van Stone, W.W. Kuldau, J.M., Engelsing, T.M.J. and Elastoff R.M. (1984). Three therapeutic communities. A prospective controlled study of narcotic addiction treatment: Process and two-year follow-up results. Arch. Gen. Psychiatry, 41, pp. 185-191.

Barr, H. (1986). Outcome of drug abuse treatment in two modalities. In: De Leon & Ziegenfuss (eds.) Therapeutic Communities for Addictions, pp. 97-108. Charles C. Thomas Publ. Springfield, Ill.

Barr, H. and D. Antes (1980). Seven years after treatment: A follow-up study of drug addicts and alcoholics treated in Eagleville Hospital's Inpatient Program. Final report to N.I.D.A. Eagleville Hosp. Research Report, Eagleville, Penn.

Bassin, A. (1973). Taming the wild para-professional. J. of Drug Issues, 3(4), pp. 353-340.

Bassin, A. (1980). Reality therapy on the therapeutic community. In: Readings of the 5th World Conference of Therapeutic Communities, Amsterdam/Noordwijkerhout, pp. 109-110. Samsom/Sijthoff, Alphen a/d Rijn.

Bastiaans, J. (1986). Isolement en bevrijding. Uitg. Balans, Amsterdam.

Berglund, G.W., Bergmark, A., Björling, B., Grönbladh, L. Lindberg, S., Oscarsson, L., Olsson, B. Segraeus, V. and C. Stensmo (1991). The Swedate Project: Interaction between treatment, client background and outcome in a one-year follow-up. J. of Substance Abuse Treatment, 8, pp. 161-169.

Biase, D.V. (1986). Education in the therapeutic community and self-esteem. The miniversity project. Paper Presented at the 10th World Conference of Therapeutic Communities, Sept. 1986, Eskilstuna.

Biase, D.V. and M. Rosenthal, (1969). Phoenix House: therapeutic communities for drug addicts. Hospital and Community Psychiatry, 20, pp. 27-30.

Biase, V. and A.P. Sullivan (1984) Succeful development of the self-concept with Therapeutic Community residents. In: Proceedings of the 8th World Conference of Therapeutic Communities, 2, 175-184. Centro Italiano di Solidarietà, Rome.

Biase, D.V., Sullivan, A.P. and B. Wheeler (1986). Daytop Miniversity - Phase 2 - College training in a Therapeutic Community - Development of self concept among drug free addict/abusers. In: De Leon & Ziegenfuss (eds.) Therapeutic Communities for Addictions.

Readings in Theory, Research and Practice. Charles C. Thomas Springfield, Ill.

Biernacki, P. (1986). Pathways from heroin addiction: Recovery without treatment. Temple University Press, Philadelphia, Penn.

Bindels, P., Stembusch, M., Tichelman, A. and R. Vernooij (1981). Onderzoeksverslag nr.1, Methadonteam-CAD, Utrecht.

Björling, B. (1986). An interdisciplinary multidimensional design for research on drugabusers in treatment and on treatment systems in Sweden. In: Therapeutic Communities, Family, Society. Proceedings of the third European Conference of Therapeutic Communities, November 1985, Bruges. Vereniging voor Opvoeding en Therapie, Gent, pp. 133-147.

Björling, B. (1989). Combined research in therapeutic communities in Sweden - Swedate Project. In: Muntarbhorn (ed.) Therapeutic communities into the 90's. Proceedings of the 11th World Conference of Therapeutic Communities, pp. 383-384. Nat. Council on Social Welfare of Thailand, Bangkok.

Blank, L., Gottsegen, M.G. and G.B. Gottsegen (eds.) (1971). Confrontation: Encounters in self and other awareness, New York.

Bowlby, J.B. (1969). Attachment and loss. Vol I, Attachment. Basic Books, New York.

Bos, C. (1977). Ouders, broers en zusters: deel van de oplossing. Driebergen Scriptie "De Horst".

Bratter, T.E. (1973). Treating alienated, unmotivated drug abusing adolescents. Am. J. of Psychotherapy, 27, pp. 585-598.

Bratter, T.E. (1978). Motivating the unmotivated. The self-help therapeutic community's biggest challenge. The Addiction Therapist, 2 & 3. pp. 84-93.

Bratter, T.E., Collabolletta, E.A., Fossbender, A.J., Pennacchia, M.C. and J.R. Rubel (1985). The American Self-Help residential Therapeutic Community. A pragmatic Treatment Approach for addicted Character-disordered Individuals. In: Bratter & Forrest (eds.). Alcoholism and Substance Abuse, pp. 461-507. The Free Press, Div. of MacMillan Inc. New York, N.Y.

Bratter, T.E. and G.G. Forrest (1985). Alcoholism and Substance Abuse Strategies for Clinical Intervention. The Free Press, Div. of MacMillan Inc. New York.

Bratter, T.E. and C.A. Hammerschlag (1975). Advocate, activist, agitator: The drug abuse program administration as a revolutionary reformer. In: Rachin & Czajowski (eds). Drug Abuse Control Administration and Politics. Lexington, Mass.

Bratter, T.E. and M. Kooyman (1981). A structured environment for heroin addicts: The experiences of a community-based American methadone clinic and a residential Dutch therapeutic community. Int. J. of Social Psychiatry, 27, pp. 189-203.

Bridger, H. (1985) Groups in open and closed systems. In: Ottenberg (ed.) The therapeutic community today. A moment of reflexion in its evolution. Proceedings of the 1st World Institute of Therapeutic Communities, pp. 54-70. Centro Italiano di Solidarietà, Rome.

Broekaert, E. (1976). De drugvrije therapeutische gemeenschap naar Synanon-model. Tijdschr. Alcohol & Drugs, 2(3), pp. 93-97.

Broekaert, E. (1981a). Inleiding tot het werk in de therapeutische gemeenschap. Uitg. European Federation of Therapeutic Commununities, VSW De Kiem Gent.

Broekaert, E. (1981b). De drugvrije therapeutische gemeenschap als orthopedagogische entiteit. Tijdschr.v. Alc. & Drugs, 7(1), pp. 16-21.

Broekaert, E. (1984). The Identity of the european therapeutic community: an adventure of therapy and education. In: Bremberg (ed.) The third generation of therapeutic communities. Proceedings of the 1st European Conference on Milieu Therapy in Eskilstuna 1982, pp. 24-27. Katrineholm.

Broekaert, E. (1990). Historische, filosofische en ideologische achtergrond van de therapeutische gemeenschap. Verslag: Kind-gezin-instelling, een werkzame driehoek. Acco, Leuven.

Brown, B.S., Buhringer, G. Kaplan, C.D. and J.J. Platt (1987). German/American report of the effective use of pressure in the treatment of drug addiction. Psychology of Addictive Behaviors 1 (1): pp. 38-54.

Bschor, F. (1986) Relevant parameters in long-term outcome of opiate addicts. In: Kaplan, Kooyman, Jansen & Smit (eds.) Proceedings of the 3rd Workshop of the European Workinggroup on Drugpolicy Oriented Research (EWODOR). Publication no.72, pp. 8-22. Institute for Preventive & Social Psychiatry, Erasmus University, Rotterdam.

Cafmeyer, I. (1986). De rol van de familieleden bij het instandhouden van de verslaving. Tijdschr. over Verslavingsproblematiek, De Sleutel, 3, pp. 4-7.

Cancrini, L., Constantini, D., Mazzoni, S., Cingolani, S. and F. Compagnoni (1985). Juvenile drug addiction: A study on typology of addicts and their families. In: Proceedings of the 9th World Conference of Therapeutic Communities, pp. 59-68. Walden House, San Francisco, Cal.

Capitanio, C., Clerici, M. Garini, R., Miragoli, F., Pisoni, G., Sangiovanni, M., Zardi, L. & Gori, E. (1985). Treatment outcome results of the therapeutic community Casa Verde. Paper presented at the 15th ICAA International Institute on the Prevention and Treatment of Drug Dependence, April 1985, Amsterdam/Noordwijkerhout.

Casriel, D. (1963). So Fair a House: The Story of Synanon. Prentice Hall Inc, Englewood Cliffs, N.Y.

Casriel, D. (1971). The Daytop story and the Casriel method. In: Blank, Gottsegen & Gottsegen (eds.). Confrontation, Encounters in Self and Interpersonal Awareness, pp. 179-193. New York.

Casriel, D. (1972). A scream away from happiness. A penetrating and detailed exploration of the principles behind scream therapy. Grosset & Dunlap, New York, N.Y.

Casriel, D. (1980). A.R.E.B.A. Accelerated re-education of emotions, behavior and attitudes. In: Readings of the 5th World Conference of Therapeutic Communities, pp. 267-269. Samsom/Sijthoff, Alphen a/d Rijn.

Casriel, D. and G. Amen (1971). Three Addicts and their Cure. Hill and Wang, New York.

Cassuto, R. (1981). Emiliehoeve 1977. Beschrijving van een drugvrije therapeutische gemeenschap vanuit sociaal psychologisch oogpunt. Katholieke Universiteit, Nijmegen. Doktoraalscriptie Sociale Psychologie.

Chinlund, S. (1978). Therapeutic community in prison. The Addiction Therapist, nr.2, 3 & 4, pp. 57-66.

Cingolani, S. (1986). Heroin addiction: Indication for family therapy on the basis of a (tentative) classification. In: Kaplan and Kooyman (eds.), Proceedings of the 15th ICAA Inst. on Prevention and Treatment of Drug Dependence, Noordwijkerhout, pp. 292-297, Inst. for Preventive and Social Psychiatry, Erasmus University, Rotterdam.

Clarck, D.M. (1984). The therapeutic community over 40 years: some personal reflections. In: Proceedings of the 8th World Conference of Therapeutic Communities, 1, pp. 15-20. Centro Italiano di Solidarietà, Roma.

Collier, W.V. (1971). An evaluation report on the therapeutic programs of Daytop Village Inc., for the period 1970-1. Research Division Daytop Village, New York, N.Y.

Collier, W.V., Hammoch, E.R. and C. Devlin (1970). An evaluation report on the therapeutic program of Daytop Village. Daytop Village Inc., New York, N.Y.

Collier, W.V. and Y.A. Hijazi (1974). A follow-up of former residents of a therapeutic community. The Int. J. of the Addictions, 9, pp. 805-826.

Comberton, J. (1982). Drugs and young people, Ward Rivers Press Ltd., Dublin.

Comberton, J. (1986). Origins and development of the therapeutic community or social learning community. Paper presented at the 10th World Conference of Therapeutic Communities, Sept. 1986, Eskilstuna.

Condelli, W.S. (1986). Client evaluations of therapeutic communities and retention. In: De Leon & Ziegenfuss (eds.), Therapeutic communities for addictions: Readings in theory, research and practice, pp. 131-139. Charles C. Thomas Springfield, Ill.

Condelli, W.S. (1989). External pressure and retention in a therapeutic community. Int. Journ. of therap. Communities, 10(4), pp. 21-33.

Coolen, J. (1985). Bondingtherapie en verslaving. Doktoraalscriptie. Katholieke Universiteit, Nijmegen.

Coombs, R.H. (1981). Back on the streets: Therapeutic communities' impact upon drug abusers. Am. J. of Drug and Alcohol Abuse, 8(2) pp. 185-201.

Costello, R. (1975). Alcoholism Treatment and Evaluation, I. and II. Int. J. Addiction, 10, pp. 251-275, 857-867.

Coulson, W.R. (1972). Groups, Gimmicks and Instant Growth. An examination of Encounter Groups and their Distortions. Harper & Row, New York.

Craig, R.J. (1984). Personality dimensions related to premature termination for am inpatient drug abuse treatment program. J. Clin. Psychol., 40, pp. 351-355.

Crawley, J.A. (1971). A case-note study of 134 out-patient drug addicts over a 17 months period. Br. J. of Addiction, pp. 209-218.

Cuskey, W.R., Richardson, A.H. and L.H. Berger (1979). Specialized therapeutic community program for female addicts. In: NIDA, Serv. Research Rep., DHEW No. (ADM) 79-880. US Gvt. Print. Off. Washington D.C.

Cutter, H.S.G., Samaraweera, A., Price, B., Haskell, D. and C. Schaeffer (1977). Prediction of treatment effectiveness in a drug free therapeutic community. The Int. J. of the Addictions 12(2-3), pp. 301-321.

De Leon, G. (1984). The therapeutic community: Study on effectiveness NIDA Services Research Monograph DHSS publication (ADM No.) 84-1286. U.S. Gvt. Printing Off., Rockville, Ma.

De Leon, G. (1985). Time in program: a decade of research. Centro Italiano die Solidarietà Roma, Italy. Proceedings of the 8th World Conference of Therapeutic Communities, pp. 115-129.

De Leon, G. (1987). Legal Pressure in therapeutic communities. Paper presented at NIDA Technical Review Meeting on Civil Commitment for Drug Abuse, Jan 26-27th 1987. U.S. Gvt. Printing Off. Rockville, Ma.

De Leon, G. (1989). Therapeutic community research: Some implications for treatment and international planning. In: Muntarbhorn (ed.) Proceedings of the 11th World Conference of Therapeutic Communities, pp. 375-382. Nat. Counc. on Soc. Welfare of Thailand, Bangkok.

De Leon, G. and G.M. Beschner (1976). The therapeutic community. In: Proceedings of the Therapeutic Communities of America Planning Conference. NIDA Service Research Report U.S. Dept. of Health, Education and Welfare, Alcohol Drug Abuse and Mental Health Administration, Rockville, Ma.

De Leon, G. and V. Biase, (1975). Encounter Group: Measurement of systolic blood pressure.Psychol. Report, 37, pp. 439-445.

De Leon, G. and D. Daily (1974). Short and longterm therapeutic communities: A follow-up and cost effectiveness comparison. The Int. J. of the Addictions, 9(5), pp. 619-636.

De Leon, G. and D. Deitch (1985). Treatment of the adolescent substance abuser in a therapeutic community. In: Friedman & Beschner (eds.). Rockville, Md. Chapter in: Treatment Services for Adolescent Substance Abusers. Treatment Monograph Series NIDA ADM 85-1342.

De Leon, G. and N. Jainchill (1981). Male and female drug abusers: Social and psychological status two years after treatment in a therapeutic community. Am. J. of Drug and Alcohol Abuse, 8(4), pp. 465-497.

De Leon, G. and N. Jainchill (1986a). Circumstances, motivation, readiness and suitability (CMRS) as correlates of treatment tenure. J. of Psychoactive Drugs, 8(3), pp. 203-208.

De Leon, G. and N. Jainchill (1986b). Can we reduce early drop-out from therapeutic communities? Paper presented at the 10th World Conference of Therapeutic Communities, Sept. 1986, Eskilstuna.

De Leon, G. and S. Schwartz (1984). The therapeutic community: What are the retention rates? Am. J. of Drug and Alcohol Abuse, 10 (2), pp. 267-284.

De Leon, G., Skodol, A. and M.S. Rosenthal (1973). Phoenix house: Changes in psychopathological signs of resident drug addicts. Arch. Gen. Psychiat. 28, pp. 131-135.

De Leon, G., Wexler, H.K. and N. Jainchill (1982). The therapeutic community: Success and improvement rates 5 years after treatment. Int. J. of the Addictions, 17, pp. 703-747.

Dederich, C. (1975). The circle and the triangle: The Synanon Social System. Synanon Foundation. Marchall, Ca.

Deitch, D.A. and J.E. Zweben (1980). Synanon: a pioneering response to drug abuse treatment and a signal for caution. In: Halpern & Levine (eds.). Proceedings of the 4th International Conference on Therapeutic Communities. Daytop Village Press, New York, N.Y. pp. 57-70

Denham, J. (1978). The British Experience. In: Verslag van het nationaal heroine symposium, 24 november 1977. De Oude Stad, Amsterdam.

Dijk, W.K. van (1971). Complexity of the Dependence Problem: Interaction of Biological Psychogenic and Sociogenic Factors in Drug Dependence. In: H.M. van Praag (ed.) Biochemical and Pharmacological Aspects of Dependence and reports on Marijuana Research, pp. 26-32. Bohn, Haarlem.

Dijk, W.K. van (1976). Alcoholisme, een veelzijdig verschijnsel. Tijdschr. v. Alc. & Drugs, 2(1), pp. 26-32.

Dijk, W.K. van (1976). Medisch model en alcoholisme. Tijdschr. v. Alc. & Drugs, 2(4), pp. 125-130.

Dijk, W.K. van (1979). De miskende alcoholist. Ned. Tijdschr. Geneesk., 123 (29), pp. 1228-1236.

Dijk, W.K. van (1980). Biological, psychogenic and sociogenic factors in Drug Dependence. In: Lettieri, Sayers & Wallenstein Pearson (eds.) Theories in Drug Abuse, selected contemporary perspectives. NIDA Monograph Series. DHHS Publ. Rockville, Ma.

Dijk-Karanika, K. van (1985). The engagement of parents and/or families in the treatment in the therapeutic community Essenlaan. In: Information and Papers. Symposium: Family, drugs and therapy, pp. 91-95. Institute for Preventive and Social Psychiatry, Erasmus University Rotterdam.

Dole, V.P. (1988). Implications of methadone maintenance for theories of narcotic addiction. J.A.M.A., 260(20) pp. 3025-3029.

Dole, V.P., Nyswander, M.E. and A. Warner (1968). Successful Treatment of 750 criminal addicts. J. of the American Medical Association, 206, pp. 2708-2711.

Dole, V.P. and M.E. Nyswander (1976). Methadone maintenance treatment. A ten year perspective. J.A.M.A., 235(19), pp. 2117-2119.

Driessen, F.M.H.M. (1987). The methadon maintenance program in Amsterdam. Some preliminary results of the analysis of the registration 1981-1984. In: Kaplan & Kooyman (eds.) Papers presented at the 15th I.C.A.A. International Institute on the Prevention and Treatment of Drug Dependence, Amsterdam/Noordwijkerhout, pp. 5-16. Institute for Preventive & Social Psychiatry, Erasmus University Rotterdam.

Edwards, G. (1979). British Policies on Opiate Addiction: Ten years working of the revised response, and options for the future. Br. J. of Psychiat., 134, pp. 1-13.

Edwards, G., Arif, A. and R. Hodgson (1981). Nomenclature and classification of drug use and alcohol related problems. Bulletin of the World Health Organization, 59, pp. 225-242.

Ellinwood, E.G., Smith, W.G. and G.E. Vaillant (1966). Narcotic addiction in males and females: A comparison. J. of the Addictions, 1, pp. 33-45.

Ellis, A. and R. Grieger (1977). Handbook of Rational-Emotive Therapy. Springer, New York, N.Y.

Endore, G. (1968). Synanon. Doubleday & Co. Inc. Garden City, New York, N.Y.

Epen, H. van (1990). Schets voor een nieuwe psychiatrie. Bohn/Stafleu/Van Loghum, Houten.

Erikson, E.H. (1963). Childhood and Society. W.W. Norton, New York, N.Y.

Fort, J.P. (1954). Heroin addiction among young men. Psychiatry, 17, pp. 251-259.

Foureman, W.C. and R. Parks (1981) The MMPI as a predictor of retention in a therapeutic community for heroin addicts. The Int. J. of the Addictions, 16(5), pp. 893-903.

Frank, R. and M. Weesie (1985). Activities of the Stichting HAD (Hulp Aan Druggebruikers) in Eindhoven for parents. Information and papers symposium: Family, Drugs and Therapy, oct. 1985, pp. 75-83. Institute for Preventive and Social Psychiatry, Erasmus University Rotterdam.

Freudenberger, H.J. (1976). The staff burn-out syndrome in alternative institutions. (Reprint from: Psychotherapy: theory, research and practice, vol.12(1), Spring 1975). The Addiction Therapist, 1(4), pp. 35-47.

Freudenberger, H.J. (1980). Burn out: The high costs of high achievement. Anchor Press/Doubleday, New York, N.Y.

Garfield, H.M. (1978). The Synanon religion. The survival morality for the 21th century. The Synanon Foundation, Marchall, Cal.

Geerlings, P. and J. de Klerk-Roscam Abbing (1985). Cathartische activerende psychodynamische groepstherapie. Tijdschrift voor Psychotherapie, 11, pp. 1-19.

Gelormino, T., Rufino, S. and L. Pappalardo (1985). Parallel family therapy. Why Parallel? Symposium: Family, Drugs and Therapy. In: Information and papers, pp. 51-67. Institute for Preventive and Social Psychiatry, Erasmus University Rotterdam.

Glaser, F.B. (1977). The origins of the drugfree therapeutic community: A retrospective history. In: Vamos & Brown (eds.) Proceedings of the 2nd World Conference on Therapeutic Communities. The Addiction Therapist, Spec. Ed., 2, 3 & 4.

Glasser, W. (1965). Reality Therapy. A new Approach to Psychiatry. Harper & Row, New York.

Gleason, D.R. (1983). SHAR House a therapeutic community working with families. Proceedings of the 7th World Conference of Therapeutic Communäities, pp. 54-56. Gateway House, Chicago.

Goffman, E. (1961). Asylums. Essays on the Social situation of mental patients and other inmates. Aldine Publishing Company, Chicago.

Gordon, A.M. (1973). Drugs and delinquency; a ten year follow-up of drug clinic patients. Br. J. Psychiatry, 142, pp. 21-25.

Gori, E., Zardi Gori, L., Clerici, M. and R.Garini (1984). 'Folow-up' da uno a sette anni di ex residenti della 'Cascina Verde', una della piu vecchie Comunita Terapeutiche italiane. In: Proceedings of the 8th World Conference of Therapeutic Communities, 2, 133-148. Centro Italiano di Solidarietà, Rome.

Gori, E., (1992). Paper presented on the follow-up research of the therapeutic community 'Cascina Verde' in Milan at the XVth World Conference of Therapeutic Communities in Venice, October 1992.

Grapendaal, M., Leuw, E. and J.M. Nelen (1991). De economie van het drugbestaan: criminaliteit als expressie van levensstijl en loopbaan. Gouda Quint, Arnhem.

Griffin, G.A. and H.F. Harlow (1966). Effects of three months of total social deprivation on social adjustment and learning in the Rhesus monkey. Child Dev. 37, pp. 535-547.

Griffith, R.R., Bigelow, G.E. and J.E. Henningfield (1980). Similarities in Animal and Human Drugtaking Behavior. In: Mello (ed.) Advances in Substance Abuse, 1, pp. 1-90. AT Press Inc., Greenwich, CT.

Gunderson, J.C., Will, O.A. and L.R. Moster (1983). Principles and Practice of Milieu Therapy. Jason Aronson, New York, N.Y.

Hadaway, P.F. (1986). Paper presented at the 15th ICAA International Institute on the Prevention and Treatment of Drug Dependence, Amsterdam /Noordwijkerhout.

Haley, J. (1971). Approaches to family therapy. In: Haley, J. (ed.) Changing families: a family therapy reader. Grune & Stratton, New York.

Haley, J. (1980). Leaving home, the therapy of disturbed young people. McGraw-Hill, New York, N.Y.

Harbin, H.T. and H.M. Maziar (1975). The families of drug abusers: a literature review. Family Process, 14, pp. 411-431.

Harlow, H.F. & M.K. Harlow (1971). Psychopathology in monkeys. In: Kimmel (ed.) Experimental Psychology. Academic Press, New York, N.Y.

Harrington, P. and T.J. Cox (1979). A twenty year follow-up of neurotic addicts in Tucson, Arizona. Am. J. of Drug and Alcohol Abuse, 6(1), pp. 25-37.

Hart, O. van der and S. Boon (1988). Schrijfopdrachten en hypnose voor de verwerking van traumatische herinneringen. Direktieve Therapie, 8, pp. 4-44.

Heckmann, W. (1986). Addicts cannot be cured by force. In: Kaplan & Kooyman (eds.) Papers presented at the 15th ICAA Int.Inst.on the Prevention and Treatment of Drug Dependence, pp. 94-100. Institute for Preventive & Social Psychiatry, Eramus University Rotterdam.

Hendriks, V. (1989). Client profile of the T.C.: who comes into treatment? In: Muntarbhorn (ed.) Therapeutic Communities into the 90's. Proceedings of the 11th World Conference of Therapeutic Communities, pp. 284-295. Nat. Council on Social Welfare of Thailand, Bangkok.

Hendriks, V.M. (1990). Addictione and psychopathology: a multidimensional approach to clinical practice. Proefschrift, Erasmus Universiteit Rotterdam.

Hill, P., Murray, R. & Thorley, A. (eds) (1979). Essentials of Postgraduate Psychiatry. Academic Press, London/New York.

Holland, S. (1983a). The effectiveness of the therapeutic community: a brief review. In: Proceedings of the 7th World Conference of Therapeutic Communities, pp. 27-33. Gateway House, Chicago.

Holland, S. (1983b). Evaluating Community-Based Treatment Programs: A model for strenghtening interferences about effectiveness. Int. J. of Therapeutic Communities, 4(4), pp. 285-305.

Holland, S. (1987). The Therapeutic Community: Treatment Outcome, Process and Client Mediators. In: Kaplan & Kooyman (eds.) Proceedings of the 15th ICAA Int. Institute on the Prevention and Treatment of Drug Dependence, pp. 57-61. Institute for Preventive & Social Psychiatry, Erasmus University, Rotterdam.

Hollidge, C. (1980). Psychodynamic aspects of the addicted personality and their treatment in the therapeutic community. In: Readings of the 5th World Conference on Therapeutic Communities. Noordwijkerhout/Amsterdam, pp. 61-86. Samsom/Sijthoff, Alphen a/d Rijn.

Hubert, M.C. and P. van Steijn (1986). Gezinstherapie binnen de ambulante laagdrempelige drugshulpverlening. In: Tijdschr. voor alc., drugs en andere psychotr. stoffen, 12, pp. 131-170.

Hunt, G.H. & Odoroff, M.E. (1962). Follow-up study of narcotic drug addicts after hospitalization. Public Health Reports, 77, pp. 44-54.

Janov, A. (1970). The primal scream: A revolutionary cure for neurosis. G.P. Putnam's Sons, New York, N.Y.

Janssen, O. and K. Swierstra (1983). On defining "hard core addicts". Instituut voor Criminologie, Groningen.

Jellinek, E.M. (1960). The disease concept of alcoholism. Hillhouse, New Brunswick.

Joe, G.W. and D.D. Simpson (1976). Relationship of patient characteristics to tenure. In: Sells & Simpson (eds.), Studies in effectiveness of treatment of drug abuse, Ballinger, Cambridge, M.A..

Jones, M. (1953). The Therapeutic Community: A new Treatment Method in Psychiatry. Basic Books, New York.

Jones, M. (1979). Therapeutic Communities: old and new. Am. J. of Drug & Alcohol Abuse 6(2), pp. 137-149.

Jones, M. (1983). Synthesis: what is social learning? In: The process of change. Routledge & Kegan Paul, London.

Jones, M. (1984). Social learning. In: The Therapeutic Community today. A moment of reflexion on its evaluation. Ottenberg (ed.), 9-44. Proceedings of the 1st World Institute of Therapeutic Communities, Castel Gandolfo, Centro Italiano di Solidarietà, Rome.

Jones, M. (1986). Democratic Therapeutic Communities (D.T.C.'s) or Programmatic Therapeutic Communities (P.T.C.'s) or both? In: De Leon & Ziegenfuss (eds.) Therapeutic Communities for Addictions, pp. 19-28. Charles C. Thomas, Springfield, Ill.

Jongerius, P.J. (1981) De psychiater als milieukundig ingenieur. Tijdschr. v. Psychiatrie, 23, pp. 317-325.

Jongerius, P.J. and R.F.A. Rylant (1989) Milieu als methode. Theorie en praktijk van de methodische milieuhantering in de GGZ. Boom, Meppel.

Jongsma, T. (1981). Wat is therapeutisch in een hiërarchisch werkende therapeutische gemeenschap? Tijdschr. v. Alcohol & Drugs, 7(3), pp. 106-110.

Jongsma, T. (1986). The psychological limits of compulsive treatment. In: Kaplan & Kooyman (eds.) Proceedings of the 15th ICAA Institute of the Prevention and Treatment of Drugdependence, pp. 103. Institute for Preventive & Social Psychiatry, Erasmus University Rotterdam.

Jongsma, T. and J.C. van der Velde (1985). Therapeutische gemeenschappen en drugsverslaafden. De mythe van de elite. Tijdschr. v. Alcohol & Drugs, 11(3), pp. 131-136.

Kalajian, B. (1979). The family re-entry group: A support system. In: Proceedings of the 4th International Conference of Therapeutic Communities, pp. 118-121. Daytop Village, New York, N.Y.

Kandel, D. (1973). Adolescent marijuana use: role of parents and peers. Science, 181, pp. 1067-1081.

Kandel, D.B., Treiman, D., Faust, R. & E. Single (1976). Adolescent involvement in legal and illegal drug use: a multiple classification analysis. Social Forces, 55, pp. 438-458.

Kantzian, E.J. (1980). An ego/self theory of substance dependence: A conceptual psychoanalytic perspective. In: Lettieri, Sayers & Wallenstein Pearson (eds.) Theories on drug abuse. Selected contemporary Perspectives, pp. 29-33. National Institute on Drug Abuse, Research monograph Series. U.S. Gvt. Printing Office, Washington, D.C.

Kaplan, C.D. and M. Wogan (1978). The psychoanalytic theory of addiction; A reevaluation by use of a statistical model. Am. J. of Psychoanalysis, 38, pp. 317-326.

Kaplan, C.D. and M. Kooyman (eds.), (1985). Proceedings of the 15th International Institute on the Prevention and Treatment of Drug Dependence of the I.C.A.A., Amsterdam/Noordwijkerhout. Institute for Preventive and Social Psychiatry, Erasmus University Rotterdam.

Kaplan, H.B., Martin, S.S. and C. Rabbin (1984). Pathways to adolescent drug use: Selfderogation, peer influence, weakening of social controls and early substance use. J. of Health and Social Behavior, 25, pp. 270-289.

Kaufman, E. (1981). Family structures of narcotic addicts. Int. J. of the Addictions, 16, pp. 273-282.

Kaufman, E. and P. Kaufmann (1979a). Multiple family therapy with drug abusers. In: Kaufman, E. & Kaufmann, P. (eds.) Family therapy of drug and alcohol abuse. Gardner Press, New York, N.Y., pp. 81-94.

Kaufman, E. and P. Kaufmann (eds) (1979b). Family therapy of drug and alcohol abuse. Gardner Press, New York, N.Y.

Kaye, A. (1987). Residents's change in the therapeutic community with regard to self-esteem, assertiveness, risk-taking and independence. Paper presented at the third conference of the European Federation of Therapeutic Communities, Dublin.

Kerr, D.H. (1986). The Therapeutic Community: A codified Concept for Training and upgrading Staffmembers working in a Residential Setting. Chapter in: De Leon & Ziegenfuss (eds.) Therapeutic Communities for Addiction, pp. 55-63. Charles C. Thomas, Springfield, Ill.

Khantzian, E.J. (1982). Psychological (structural) vulnerabilities and the specific appeal of narcotics. Annals of the New York Academy of Science, 398, pp. 24-32.

Klagsburn, M. and D.L. Davis (1977). Substance abuse and family interaction. Family Process, 16, pp. 149-173.

Kolk, B.A. van der (1987). The separation cry and the trauma response. In: B.A. van der Kolk, Psychological Trauma, pp. 31-62. American Psychiatric Association, Washington, D.C.

Kolk, B.A. van der (1992). The re-enactment of the trauma in psychotherapy. Paper presented at the World Conference of the Int. Soc. for Traumatic Stress Studies Trauma and Tragedy, june 1992, Amsterdam.

Kooyman, I.G.C. (1992). Emiliehoeve. Scriptie oriëntatiestage H.S.A., Den Haag.

Kooyman, M. (1975a). From Chaos to a Structured Therapeutic Community: Treatment Program on Emiliehoeve. A Farm for Young Addicts. Bulletin on Narcotics, 27(1), pp. 19-26.

Kooyman, M. (1975b). Problematiek rondom opname en ontslag in de psychiatrie. Goed, beter worden. Jubileumuitgave, pp. 25-29. KMMG, Utrecht.

Kooyman, M. (1975c). Een nieuw behandelbeleid roept vragen op. Tijdschr. v. Alcohol & Drugs, 1, pp. 63-65.

Kooyman, M. (1975d). Drie jaar Emiliehoeve. Tijdschr. v. Alcohol & Drugs, 1, pp. 3-8.

Kooyman, M. (1977). Gedwongen opname, een zinvol hulpmiddel bij de behandeling van verslaving? Medisch Contact (32), pp. 1379-1382.

Kooyman, M. (1978). Pathology in the therapeutic community: The role of the psychiatrist. In: Corelli (ed.) Proceedings of the third World Conference of Therapeutic Communities, pp. 263-267. Centro Italiano di Solidarietà, Rome.

Kooyman, M. (1978). The history of the therapeutic community movement in Europe. In: Vamos & Brown (eds.) Proceedings of the 2nd World Conference of Therapeutic Communities, Montreal, 1977. The Addiction Therapist, Special Edition, 2, 3 & 4, pp. 29-32.

Kooyman, M. (1980). De drugvrije behandeling van verslaafden. Heroineverslaving, Boerhaave cursus, pp. 123-129. Faculteit der Geneeskunde, Rijksuniversiteit Leiden.

Kooyman, M. (1981). Nieuwe initiatieven bij de krisisopvang. Verslag van een studiedag: De verslaafde en zijn zorg, pp. 11-14. T.G. de Kiem, Oosterzele.

Kooyman, M. (1982). De professionele hulpverlener in de therapeutische gemeenschap voor verslaafden. Tijdschr. v. Alcohol & Drugs, 8(2), pp. 61-64.

Kooyman, M. (1983). Driehoeken en cirkels. De behandeling in de drugvrije therapeutische gemeenschap. Voordracht gehouden op symposium ter gelegenheid van het 5-jarig bestaan van Welland, Heerlen.

Kooyman, M. (1984a). The drug problem in The Netherlands. J. of Subst. Drugabuse Treatment, 1, pp. 125-130.

Kooyman, M. (1984b). The history of the therapeutic community movement in Europe. In: Vamos & Brown (eds.) Proceedings of the 2nd World Conference of Therapeutic Communities. Portage Press, Montreal, pp. 29-32.

Kooyman, M. (1984c). Naar een consequent heroinebeleid. Tijdschr. v. Alcohol & Drugs, 19(4), pp. 160-163.

Kooyman, M. (1985a). Acting out behavior in the therapeutic community. In: Ottenberg (ed.) The therapeutic community today. A moment of reflection on its evaluation. Proceedings of the 1st World Institute of Therapeutic Communities. Centro Italiano di Solidarietà, Roma.

Kooyman, M. (1985b). De therapeutische gemeenschap als psychotherapeutisch instrument bij de behandeling van drugverslaafden. Tijdschr. v. Psychiatrie, 27(3), pp. 160-179.

Kooyman, M., (1985c). Follow-up of former residents of the Emiliehoeve, therapeutic community for drug addicts. In: Proceedings of the 34th ICAA International Congress on Alcoholism and Drug Dependence, pp. 231-234. AADAC, Calgary.

Kooyman, M. (1986a). The psychodynamics of therapeutic communities for treatment of heroin addicts. In: De Leon & Ziegenfuss (eds.) Therapeutic communities for Addictions. Readings in Theory, Research and Practice. Charles C. Thomas, Springfield, Ill.

Kooyman, M. (1986b). Die Entwicklung der Behandlungsmasznahmen für Drogen-abhängige in der Niederlanden. Wiener Zeitschr. für Suchtforschung, 9(1-2), pp. 51-54.

Kooyman, M. (1986 c). Een terugblik op 17 jaar behandeling van drugsverslaafden. Tijdschr. voor alc. drugs en andere psychotr. stoffen, 12, nr.2, pp. 61-66.

Kooyman, M. (1986 d). De intramurale drugshulpverlening in de verdrukking. Marge, 3, Welzijnsmaandblad, 4, Jrg. 40, 35-39.

Kooyman, M. (1987). Report on the innovation approach of the Centro Italiano di Solidarietà in Rome to the drug problem in Italy. In: Demand reduction in practice. A worldwide review of innovative approaches. WHO/I.C.A.A., Lausanne.

Kooyman, M. (1989). Therapeutic communities in East West perspective. In: Muntarbhorn (ed.) Therapeutic Communities into the 90's. Proceedings of the 11th World Conference of Therapeutic Communities, pp. 105-106. National Council on Social Welfare of Thailand, Bangkok.

Kooyman, M. (1991) Tederheidstekort en verslaving. In: de Groot en Kruijt (red.), Tederheid, pp. 89-99. S.I.G.O., Boom, Meppel.

Kooyman, M. (1992). The Therapeutic Community for Addicts; Intimacy, Parent Involvement and Treatment Outcome. Dissertation. Universiteitsdrukkerij, Erasmus Universiteit Rotterdam.

Kooyman, M. and T.E. Bratter (1980). De noodzaak van confrontatie en structuur bij de behandeling van drugverslaafden. Tijdschr. v. Alcohol & Drugs, 6(1), pp. 27-33.

Kooyman, M. and M.C.M. Esseveld (1984). Encountergroepen en andere groepstherapieën bij verslaafden. In: Schwencke (ed.) Groep en systeem. Nederlandse Vereniging voor Groepstherapie, Bergen op Zoom.

Kooyman, M. and C.D. Kaplan (1986). The position of research in the therapeutic community today. In: Proceedings of the 3rd European Conference of Therapeutic Communities, Bruges, pp. 130-132. Vereniging voor Opvoeding en Therapie, Gent.

Kooyman, M., Lugt, B. van der, and M. Sonnen (1979). De praktijk van de detoxificatie van de heroine verslaafde. Klinische les. Ned. Tijdschr. Geneeskunde, 123(27), pp. 1137-1140.

Kooyman, M. and P. van Steijn (1990). The Family and the Use of Illegal Drugs. The Role of the Family and their Primary Social Groups in Treatment and Prevention in Europe. Report of a WHO study. EUR/HFA, target 14. WHO, Regional Office for Europe, Kopenhagen.

Kraemer, G.W. (1985). Effects of differences in early social experiences on primate neurological-behavioral development. In: Reite, Fields, Orlando (eds.) The psychology of Attachment and Separation. Academic Press, New York, N.Y.

Krystal, H. (1988). Integration and self-healing. Affect, trauma, alexithymia. The analytic Press, L. Erlbaum Ass. Hillsdale N.J..

Krystal, H (1970). Drug Dependence; aspects of ego function. Wayne State University Press, Detroit.

Kunz, D., and H. Kampe (1985). Zum Problem des Therapieabbruchs von Heroin-abhängigen. Suchtgefahren 31, pp. 146-154.

Kramer, J.C. (1972). The place of civil commitment in the management of drug abuse. In: Harris et al. (eds.) Drug dependence. Univ. of Texas Press, Austin.

Lakoff, R. (1978). Psychopathology within a therapeutic community: A review of psychiatric consultations from a drug addiction program. The Addiction Therapist, 2, 3 & 4, pp. 48-51.

Lakoff, F. (1984). Integrating the family into the therapeutic community concept. Proceedings of the 8th World Conference of Therapeutic Communities, 1, pp. 473-476. Centro Italiano di Solidarietà, Rome.

Lanini, Y. (1985). Research on the characteristics of the family of addicts in treatment in a therapeutic community. Therapeutic community, family, society. Proceedings of the 3rd European Conference of Therapeutic Communities, pp. 67-70. Vereniging voor Opvoeding en Therapie V.Z.W., Gent.

Laudermilk, W. (1981). Staff training in the therapeutic community. In: Proceedings of the 6th World Conference of Therapeutic Communities, pp. 77-78. DARE, Astrad Prod. House, Manila.

Lutterjohann, M. (1984). Rational Emotive Therapy with Addicts Changes and Limitations of Application in Therapeutic Communities. In: L. Bremberg (ed.) Proceedings of the 1st European Conference on Milieu Therapy, Sept. 1982, Eskilstuna, pp. 109-116. Katrineholm.

Maertens, J. (1982). De drugvrije therapeutische gemeenschap "De Sleutel". Tijdschr. v. Alcohol & Drugs, 8, nr. 1, pp. 27-31.

Maertens, J. (1986). Integratie van client-centered therapie en bonding therapie in een residentiële setting. Tijdschr. over Verslavingsproblematiek, De Sleutel, 3(1) pp. 9-12.

Mahon, T. (1973). Therapy or brainwashing? Drugs & Society, 2(5) pp. 7-10.

Majoor, B. (1986). What is burn out? In: Kaplan & Kooyman (eds.) Proceedings of the ICAA 15th International Institute on Prevention and Treatment of Drug Dependence, Amsterdam/Noordwijkerhout, pp. 150-155. Institute for Preventive & Social Psychiatry, Erasmus University Rotterdam.

Maloney, M. (1985). The challenge of growth and change. Daytop Village Family Association. In: Bridging Services. Proceedings of the 9th World Conference of Therapeutic Communities, pp. 244. Walden House, San Francisco, Ca.

Marlatt, G.A. (1985). Relapse Prevention: theoretical, rational and overview of the model. In: Marlatt & Gordon (eds.) New York Relapse Prevention. Guilford.

Maslow, A.H. (1967). Synanon and eupsychia. J. of Humanistic Psychology, 7, pp. 28-35.

Maslow, A.H. (1968). Toward a Psychology of Being. Van Nostrand Comp. Princeton N.J.

Mason, W.A. (1967). Motivational aspects of social responsiveness in young chimpanzees. In: Stevenson (ed.) Early behaviour, comparative and developmental aspects. H. John Wiley & Sons, New York, N.Y.

McLellan, A.T. et al. (1984). The psychiatrically severe drug abuse patient: Methadone maintenance or therapeutic community?. Am. J. of Drug and Alcohol Abuse, 10(1), pp. 77-95.

McLellan, A.T., Luborsky, L., Woody, G.E. and C.P. O' Brien (1980). An improved diagnostic evaluation instrument for substance abuse patients. The Addiction Severity Index. J. of Nerv. and Ment. Disease, 168, pp. 26-33.

Meer, Chr. van der (1985). Possibillities and limitations of parent involvement in relation to the treatment of drug abusers in a therapeutic commune. In: Information and papers. Symposium: Family, Drugs and Therapy, pp. 75-83. Institute for Preventive & Social Psychiatry, Erasmus University Rotterdam.

Meer, Chr. van der (1986). Parent involvement in a drugfree therapeutic community (Emiliehoeve). In: Kaplan & Kooyman (eds.) Proceedings of the I.C.A.A. 15th International Conference on the Prevention and Treatment of Drug Dependence, Amsterdam/Noordwijkerhout, pp. 316-319. Institute for Preventive & Social Psychiatry, Erasmus University Rotterdam.

Minuchin, S. (1974). Families and Family therapy. Harvard University Press, Cambridge, Mass.

Moreno, J.L. (1959). Psychodrama. Vol. I, II, III. Beacon Press, New York, N.Y.

Moreno, J.L. (1969). The Viennese origins of the encounter movement, paring the way for existentialism, group psychotherapy and psychodrama. Group Psychotherapy, 22, pp. 8-9.

Mowrer, O.H. (1959). The crisis in psychiatry and religion. Van Nostrand Comp., Princeton, N.J.

Mowrer, O.H. (1977). Therapeutic groups and communities in retrospect. In: Proceedings of the first World Conference of Therapeutic Communities, Sept. 1976, Norköping, The Portage Press, Montreal.

Mullan, M. (1989). Alcoholism and the new genetics. Br. J. Addiction, 12, pp. 1433-1440.

Nabitz, U. & Hermanides, R. (1986). Reducing Drop-out Rates. A Treatment Evaluation Study in the Jellinekcentrum. In Kaplan, Kooyman, Jansen & Smit (eds) Proceedings of the 3rd Workshop of the European Workinggroup on Drugpolicy Oriented Research (EWODOR), pp. 23-33. Publ. nr. 72, Institute for Preventive & Social Psychiatry, Erasmus University Rotterdam.

NIDA (1982). Data from Client Oriented Data Aquisition Program (CODAP), NIDA, Rockville, Ma.

Noach, E.L. (1980). Farmacologie van de opiaten. In: Heroine verslaving. Boerhaave cursus, pp. 25-33. Faculteit der Geneeskunde, Rijksuniversiteit Leiden.

O'Brien, W.B. (1981). Transcultural Aspects of Therapeutic Communities In: The therapeutic community in various cultures worldwide. Proceedings of the 6th World Conference of Therapeutic Communities, pp. 16-18. Astrad Prod. House, Manila.

O'Brien, W.B. (1983). The family and its interaction with the TC: a world wide perspective. In: Proceedings of the 7th World Conference of Therapeutic Communities, pp. 19-21. Gateway House, Chicago.

Ogborne, A.C. and C. Melotte, (1977). An evaluation of a therapeutic community for former drugs users. Brit. J. of Addiction, 72, pp. 75-82.

Osterhues, U.J. (1990). Die Behandlung borderline-kranker Suchtpatienten in Therapeutischen Gemeinschaften. Drogen-Report-Arbeitsskript Nr. 1/90, Druck, Rumpel, Nürnberg.

Ottenberg, D.J. (1976). The process of recovery in the therapeutic community. In: Vamos & Devlin (eds.) Proceedings of the 1st World Conference on Therapeutic Communities in Norköping. Portage Press, Montreal.

Ottenberg, D.J. (1978). Responsible concern. The Addiction Therapist, 3(1) pp. 67-68.

Ottenberg, D.J. (1984). Therapeutic community and the danger of the cult phenomenon. In: Bremberg, L. (ed.) Third generation of therapeutic communities. Proceedings of the 1st Eur. Conference on Milieutherapy, Sept. 1982, Eskilstuna, pp. 218-238. Katrineholm.

Ottenberg, D.J. (1988). An Italian program of parallel family involvement in the treatment of drug dependence. Centro Italiano di Solidarietà, Rome.

Ottenberg, D.J. (1990). The educational value system of therapeutic communities. In: Hellinckx, Broekaert, Vandenberge & Colton (eds.) Innovations in Residential Care, pp. 135-142. ACCO, Leuven.

Ottenberg, M. (1986). Staff Issues. In: Kaplan & Kooyman (eds.) Proceedings of the ICAA 15th International Institute on Prevention and Treatment of Drug Dependence, Amsterdam/Noordwijkerkhout. pp. 148-150, Institute for Preventive and Social Psychiatry, Erasmus University, Rotterdam.

Palazzoli, M. Selvini-, Boscolo, L., Cecchin, G. and G. Prata (1978). Paradox and counterparadox. Jason Aronson, New York, N.Y.

Parsons, T. (1951). The social system. Glencoe, The Free Press, New York, N.Y.

Paschelke, G., Philipsen, P. & Kremer, H. (1986). The myth of compulsion in longterm clinical rehabilitation treatment of young delinquent drug abusers. In: Kaplan & Kooyman (eds.) Proceedings of the 15th ICAA, Institute on the Prevention and Treatment of Drug Dependence, Amsterdam/Noordwijkerhout, pp. 99-102. Institute for Preventive & Social Psychiatry, Erasmus University Rotterdam.

Patton, T. (1973). The Synanon Philosophy. Synanon Research Institute, Marchall, Cal.

Pescor, M.J. (1943). Follow-up study of treated narcotic drug addicts. U.S. Gvt. Printing Office, Washington. Public Health Reports, Suppl. nr. 170.

Pesso, A. (1980). Psychomotor/Pesso system psychotherapy. In: Readings of the 5th World Conference of Therapeutic Communities, Noordwijkerhout, pp. 111-128. Samsom/Sijthoff, Alphen a/d Rijn.

Petersen, D.M. (1974). Some reflections on compulsary treatment of addiction. In: Inciardy & Chambers (eds),Drugs and the Criminal Justice System. SAGE Publ., Beverly Hills.

Piperopoulos, G.P. (1977). The 'Concept' as a Response to the Turbulant sixties. The Addiction Therapist 2(2) pp. 62-65.

Platt, J.J. (1986). Heroine addiction. Theory, research and treatment. R.E. Krieger Publ., Malabar, Fl.

Platt, J.J. (1975). "Addiction-proneness" and personality in heroin addicts. J. of Abnormal Psychology, (84), pp. 303-306.

Pompi, K. and J. Resnick (1987). Retention in a therapeutic community for court referred adolescents and young adults. Am. J. of Drug and Alcohol Abuse, pp. 309-325.

Pompi, K.F., Shreiner, S.C. and J.L. McKey (1979). Abraxas: a first look at outcome. Abraxas Foundation, Pittsburgh, P.A.

Preston, C.A. and L.L. Viney (1984). Self- and Ideal Selfperception of Drug Addicts in Therapeutic Communities. The Int. J. of the Addictions, 19(7) pp. 805-818.

Rack, S. (1957). Narcotic Bondage: a general theory of dependence on narcotic drugs. Am. J. of Psychiatry, 114, pp. 165-170.

Rado, S. (1933). The psychoanalysis of pharmacothymia (drug addiction). Psychoanalytic Quarterly, 2, pp. 1-23.

Ramirez, E. (1973). Phoenix House: a therapeutic community program for the treatment of drug abusers and drug addicts. Yearbook of Drug Abuse. Bril & Harms, New York.

Ravndal, E. and P. Vaglum (1991). Psychopathology and substance abuse as predictor of program completion in a therapeutic community for drug abusers: a prospective study. Acta Psychiat. Scand. 83, pp. 217-222.

Ravndal, E. and P. Vaglum (1992). H.I.V. drug abusers in a hierarchical therapeutic community. A prospective study. Nordic J. of Psychiatry (in press).

Ree, J.M. and H.M. Fraenkel (1987). Neuropeptides and drug dependence. In: Kaplan & Kooyman (eds.) Proceedings of the 15th ICAA Institute on the Prevention and Treatment of Drug Dependence, Amsterdam/Noordwijkerhout, pp. 209. Institute for Preventive and Social Psychiatry, Erasmus University, Rotterdam.

Reilly, D.M. (1976). Family factors in the etiology and treatment of youthfull drug abuse. Family Therapy, 2, pp. 149-171.

Rogers, C.R. (1970). Encounter groups. Pelican Books, Harmondsworth, Middlesex, Penguin Press, Harper and Row, New York, N.Y..

Roland, S. (1986). The therapeutic community: treatment outcome, process and client mediators.

Rose, M., Battjes, R., and C. Leukefeld (1984). Family life skills. Training for drug abuse prevention. In: NIDA, Dir. of Clin. Research. DHHS Publ.no (ADM) 84-1340. U.S. Gvt. Print. Off., Washington D.C.

Rosenthal, M.S (1977). Paper presented at the NIOV (Netherlands Institute for the Training of Addiction Therapists), The Hague.

Rosenthal, M.S. (1984). Therapeutic Communities: A treatment alternative for many but not all. J. of Substance Abuse Treatment, 1, pp. 55-58.

Rosenthal, M.S. and D.V. Biase (1969). Phoenix Houses, therapeutic communities for drugaddicts. Hospital and Community Psychiatry, 20, 26-38.

Rosenthal, M.S. and D.V. Biase (1972). Phoenix Houses: Therapeutic Communities for Drug Addicts. In: Kaplan & Sadock (eds.) Groups and drugs. Jason Aronson, New York.

Sachs, J.G. and N. Levy (1979). Objective personality changes in residents of a therapeutic community. Am. J. of Psychiatry, 136(6) pp. 796-799.

Salmon, R. & S. Salmon (1977). The causes of heroin addiction. A review of the literature. Part II. Int. J. of the Addictions, 12, pp. 937-951.

Sansone, J. (1980). Retention patterns in a therapeutic community for the treatment of drug abuse. The Int. J. of the Addictions, 15(5) pp. 711-736.

Savage, L.J. and D.D. Simpson (1978). Illicit drug use and return to treatment: follow-up study of treatment admissions to DARP during 1969-1971. Am. J. Drug & Alcohol Abuse 5(1), pp. 23-38.

Schaap, G.E. (1977). Hoog-Hullen: van sanatorium naar drugvrije therapeutische gemeenschap. Tijdschr. v. Alcohol & Drugs, 7, pp. 106-110.

Schaap, G.E. (1978). A New Dutch Experiment: Hoog Hullen, a Drugfree Therapeutic Community for Alcohol Addicts. The Addiction Therapist, 2, 3 & 4. pp. 170-180.

Schaap, G.E. (1980). Democratic and Concept based Therapeutic Communities. In: Readings of the 5th World Conference of Therapeutic Communities, pp. 158-165. Samsom/Sijthoff, Alphen a/d Rijn.

Schaap, G.E. (1981). De hiërarchisch gestructureerde therapeutische gemeen schap. Een (nog) vreemde eend in de bijt van de therapeutische gemeenschappen. MGV 11, pp. 970-986.

Schaap, G.E. (1985). Nieuwe ontwikkelingen in diagnostiek en behandeling van verslaafden. In: Van Berkestijn, Giel & Van den Hoofdakker (eds.) Blokken van de puzzel: over klinische psychiatrie en haar grensgebieden. Uitg. PUK, Groningen.

Schaap, G.E. (1987). De therapeutische gemeenschap voor alcoholisten: diagnostiek behandeling en effectiviteit bij afhankelijkheidsproblemen. Van Gorcum, Assen.

Schippers, G.M. (1984). Wat verstaan we onder "probleem drinken", alcoholmisbruik en alcoholverslaving. In: Herkenning en behandeling van alcoholmisbruik, Boerhaave cursus, pp. 1-12. Rijksuniversiteit, Leiden.

Scholer, Y.A.R.W. (1986). Legal compulsion in rehabilitation: conductive or detrimental. In: Kaplan & Kooyman (eds.) Proceedings of the 15th ICAA, Institute on the Prevention and Treatment of Drug Dependence, Amsterdam/Noordwijkerhout, pp. 96-98. Institute for Preventive & Social Psychiatry, Erasmus University, Rotterdam.

Schutz, W. (1975). Grondbeginselen van encounter. Samsom, Alphen a/d Rijn.

Segraeus, V. (1985). Treatment ideology and practice among Swedish TC personnel. Paper presented at the third European Conference of Therapeutic Communities, November 1985, Bruges.

Segraeus, V. (1992). Outcome results of therapeutic communities in the Swedate Project. Paper presented at the 8th Workshop of the European Workinggroup on Drugpolicy Oriented Research (EWODOR), May 1992, Rome.

Seldin, N.E. (1972). The family of the addict: A review of the literature. The Int. J. of the Addictions, 7(1) pp. 97-107.

Sells, S. and D. Simpson (1980). The case for drug abuse treatment effectiveness based on the DARP research program. Br. J. of Addictions, 75, pp. 117-131.

Shelly, J. and A. Bassin (1965). Daytop lodge: a new treatment approach for drug addicts. Corrective Psychiatry 11, pp. 186-195.

Shostrun, E.L. (1967). Man, the Manipulator. The inner Journey from Manipulation to Actualization. Abingdon Press, Nashville.

Sijlbing, G. (1981). Methadon, GG en GD, Amsterdam. SWOAD, Amsterdam.

Simon, S.I. (1974). The Synanon Game. Synanon University Press, San Francisco, Cal.

Simpson, D.D. (1979). The relation of time spent in drug abuse treatment to posttreatment outcome. Am. J. Psychiatry, 136, pp. 1449-1453.

Simpson, D.D. (1986). 12-year Follow-up outcomes of opioid addicts treated in therapeutic communities. In: De Leon & Ziegenfuss (eds.) Therapeutic Communities for Addictions. Charles C. Thomas Publ. Springfield, Ill., pp. 109-120.

Simpson, D.D. and M.R. Lloyd (1978). Alcohol and Illicid Drug Use: Follow-up study of treatment admissions to DARP, during 1969-1971. Am. J. Drug Alc. Abuse, 5(1) pp. 1-22.

Simpson, D.D. and M.R. Lloyd (1979). Client evaluations of drug abuse treatment in relation to follow-up outcomes. Am.J. of Drug & Alcohol Abuse 6(4), pp. 397-411.

Simpson, D.D. and M.R. Lloyd (1981). Alcohol use by opioid addicts during a four year follow-up after drug abuse treatment. J. Stud. Alcohol, 42, pp. 323-335.

Simpson, D.D., Savage, L.J., Lloyd, M.R. and S.B. Sells (1978). Evaluation of Drug Abuse Treatments based on First Year Follow-up. NIDA Research Monograph. U.S. Gvt. Printing Office, Washington D.C.

Simpson, D.D. and S.B. Sells (1981). Highlights of the DARP Follow-up Research on the Evaluation of Drug Abuse Treatment Effectiveness. NIDA Monograph. NIDA, Washington D.C.

Simpson, D.D. and S.B. Sells (1982). Evaluation of drugabuse treatment effectiveness: Summary of the DAPR follow-up research. In: NIDA Treatment research report, Washington, D.C.

Stanton, A.H. and M.S. Schwartz (1954). The mental hospital. Tavistock Publ., London.

Stanton, M.D. (1979). Drugs and the family: A review of the recent literature. Marriage and Family Review, 2, pp. 1-10.

Stanton, M.D. (1980). Some overlooked aspects of the family and drug abuse. In: Ellis, B. (ed.) Drug abuse from the family perspective: coping in a family. U.S. Gvt. Printing Office, Washington, D.C.

Stanton, M.D. (1983). Drug abuse in the family. In: Proceedings of the 7th World Conference of Therapeutic Communities, pp. 21-22. Gateway House, Chicago.

Stanton, M.D. and T.C. Todd (1982). The family therapy of drug abuse and addiction. Guilford Press, New York,

Steffenhagen, R.A. (1980). Self-esteem theory of drug abuse. In: Lettieri, D.J., Sayers, M. and H.W. Parson (eds.) Theories on Drug Abuse: Selected contemporary perpspectives. U.S. Gvt. Printing Office, Washington, D.C. DHHS Pub. no. (ADM) 80-967.

Steijn, P. van and M.C. Hubert (1986). Gezinstherapie binnen de laagdrempelige ambulante drug hulpverlening. Tijdschr. v. Alcohol & Drugs, 4, pp. 147-150.

Stensmo, C. (1989). Aspect on process and outcome concerning Swedish T.C.'s. In: Muntharbhorn (ed.) Therapeutic communities in the 90's. Proceedings of the 11th World Conference of Therapeutic Communities, pp. 358-367. National Council of Social Welfare of Thailand, Bangkok.

Stoop, B., Bohmers, W. and K. Hofman (1987). Introductie van een gedragsverslavingsmodel bij het stoppen en verminderen van het gebruik van drugs. Tijdschr. v. Alc. & Drugs, 13(2) pp. 56-58.

Sugarman, B. (1974). Daytop Village, a therapeutic community. Holt, Rinehart & Winston, New York.

Sullivan, M.S. (1937). Socio-psychiatric research. Am. J. of Psychiatry, 87, pp. 989.

Suomi, S.J. (1984). The development of affect in Rhesus monkeys. In: Fox, Davidson, Lawrence (eds.) The psychology of affective development, pp. 119-159. Erlbaum Ass., Hillsdale, N.Y.

Suomi, S.J. and H.F. Harlow (1972). Social rehabilitation of isolate reared monkeys. Dev. Psychology, 6, pp. 487-496.

Suomi, S.J., Harlow, H.F. and C.J. Domek (1970). Effects of repetitive mother-infant separation of young monkeys. J. of Abnorm. Psychology, 76, pp. 161-172.

Swierstra, K. (1986). Young Dutch heroin users in the Eighties: divergence in pattern. In: Kaplan & Kooyman, (eds.) Proceedings of the 15th ICAA Institute on Prevention and

Treatment of Drug Dependence, Amsterdam/Noordwijkerhout, pp. 194-198. Institute for Preventive & Social Psychiatry, Erasmus University, Rotterdam.

Swierstra, K. (1987). Heroïne verslaving: levenslang of gaat het vanzelf over? Tijdschr. v. Alcohol & Drugs., 13, pp. 17-92.

Szasz, T.S. (1958). The role of the counter-phobic mechanism in addiction. J. Am. Psychoanal. Assoc., 6, pp. 309-325.

Szasz, T.S. (1975). Ceremonial chemistry: Ritual Persecution of Drug Addicts and Pushers. Doubleday, Garden City, N.Y.

Szasz, T.S. (1970). The manufacture of madness: A cooperative study of the Inquisition and the Mental Health Movement. Harper & Row, New York.

Tarbell, L.A. (1985). The young group. Bridging Services. Proceedings of the 9th World Conference of Therapeutic Communities, pp. 245-246. Walden House, San Francisco, Ca.

Tennant, C. and E. Bernardi (1988). Childhood loss in alcoholics and narcotic addicts. Br. J. of Addiction, 83, pp. 695-703.

Trimbos, C.J.B.J. (1980). Alcoholisme: een riskante vorm van deviant gedrag. In: Querido & Roos (eds.) Controversen in de Geneeskunde I. Bunge, Utrecht.

Uchtenhagen, A. (1985a). Drug abuse treatment and program evaluation in Europe. In: Baro, Casselman & Pele (eds.) Drug related problems in Europe. Proceedings of a WHO workshop on the prevention and treatment of drugdependence, Brussels, 1983. pp. 79-144. ACCO, Leuven.

Uchtenhagen, A. (1985b). Behavior: Personal and Social Controls. In: Ottenberg, (ed.) The therapeutic community today. A moment of reflection on its evolution. Proceedings of the 1st World Institute of Therapeutic Communities. Centro Italiano di Solidarietà, Rome.

Uchtenhagen, A. and D. Zimmer-Höfler (1981). Theoretisches Modell zur Unterpretation devianter Karrieren. In: Tongue, E. (ed.). Papers presented at the 11th ICAA International Institute on the Prevention and Treatment of Drug Dependence, Vienna. ICAA Publ., Lausanne.

Vaglum, P. & Fossheim, I. (1980). The results of different institutional treatment program; are they different in different groups of drug abusers? Acta Psychiatrica Scandinavica, 62, pp. 21-28.

Vaillant, G.E. (1966). A 12-year follow-up of New York narcotic addicts: I. The relation of treatment to outcome. Am. J. Psychiatry, 122, pp. 727-737.

Vaillant, G.E (1966). A 12-year follow-up of New York narcotic addicts: III Some social and psychiatric characteristics. Archives of General Psychiatry, 15, pp. 599-609.

Vaillant, G.E. (1973). A 20-year follow-up of New York narcotic addicts. Archives of General Psychiatry, 29, pp. 273-241.

Vamos, P. and J.J. Devlin (1975). The evolution of the addiction therapist: a training imperative. The Addiction Therapist, 1(2) pp. 27-33.

Vandenbroele, H. (1991). Pre-treatment and Dropping-out from a Therapeutic Community. Paper presented at the 7th Workshop of the European Working group on Drug-policy Oriented Research, EWODOR, Palermo.

Vandenbroele, H., Boffin, N., Veldeman, L. & Deun, P. van (1989). Evaluatie-onderzoek in een drugvrije therapeutische gemeenschap: resultaten en feedback van een follow-up onderzoek. Tijdschr. v. Alcohol & Drugs, 15(2) pp. 63-77.

Velde, J.C. van de, Schaap, G.E. & Land, H. (1989). Effectiviteit van klinische behandeling van alcoholisten. Tijdschr. v. Psychiatrie, 31(2) pp. 62-80.

Vos, H. (1984). The Role of the Psychotherapist in the Re-entry Program. J. of Psychoactive Drugs, 16(11) pp. 91-93.

Wagenaar, J. (1981). New drug programmes fashioned after therapeutic communities: a clinical introduction center. In: Proceedings of the 6th World Conference of Therapeutic Communities, pp. 78-79. DARE, Astrad Prod. House, Manila.

Walburg, J.A. (1984). Voorkomen en behandeling van recidieven. In: Herkenning en behandeling van alcoholmisbruik. Boerhaave cursus. Rijksuniversiteit, Leiden.

Waldorf, D. (1971). Social controll in therapeutic communities for the treatment of drug addicts. The Int. J. of the Addictions, 6(1) pp. 29-43.

Waldorf, D. and P. Biernacki (1981). The natural recovery from opiate addiction. Some preliminary findings. J. Drug Issues, 1, pp. 61-74.

Weber, G.H. (1957). Conflicts between professionals and non-professional personnel in institutional delinquency treatment. J. of Crim. Law, Criminology and Police Science, 48, pp. 28-30.

Weber, M. (1958). Essays in sociology. Translated, edited and with an introduction by Gerth & Wright Mills. Oxford University Press, New York, N.Y.

Wecks, J.R. (1962). Experimental Morphine Addiction: Method for automatic Intravenous Injections in Unrestrained Rats. Science, 138, pp. 143-144.

Wexler, H.K. (1986). Therapeutic Communities within prisons. In: De Leon & Ziegenfuss (eds.) Therapeutic Communities for Addictions, pp. 227-237. Charles C. Thomas, Springfield, Ill.

Wexler, H.K. and G. de Leon (1977). The therapeutic community: Multivariate prediction of retention. Am. J. of Drug and Alcohol Abuse, 4(2) pp. 145-151.

Wilson, S. (1978). The effect of treatment in a therapeutic community on intravenous drug abuse. Br. J. of Addiction, 73 (4) pp. 407-411.

Wilson, S.R. and B.M. Mandelbroke (1978). The relationship between duration of treatment in a therapeutic community for drug abusers and subsequent criminality. J. Med. Psychology, 132, pp. 487-491.

Wilson, S.R. and B.M. Mandelbroke (1985). Reconviction rates of drug dependent patients treated in a residential therapeutic community: a 10-year follow-up. Br. Med. Journal, 291, pp. 105.

Winick, C. (1962). Maturing out of narcotic addiction. Bulletin on Narcotics, 14, pp. 1-7.

Winick, C. (1980). An empirical assessment of therapeutic communities in New York City. In: Brill & Winick (eds.) The Yearbook of Substance Use and Abuse, vol.II, pp. 251-285. Human Sciences Press, New York, N.Y.

Wolfe, A. & Schwartz, E.K. (1962). Psychoanalysis in Groups. Grune & Stratton, New York.

Wurmser, L. (1972). Drug abuse nemesis of psychiatry. American Scholar, 41, pp. 393-397.

Wijngaard, G.F. van de (1987). Rethinking the qualities of methadone. In: Kaplan & Kooyman (eds.) Proceedings of the 15th I.C.A.A. International Institute on the Prevention and Treatment of Drug Dependence. Amsterdam/Noordwijkerhout. pp. 173-179.

Institute for Preventive & Social Psychiatry, Erasmus University, Rotterdam.

Yablonski, L. (1967). Synanon, the Tunnel back. MacMillan, New York.

Yablonski, L. (1990). Therapeutic community; a succesful approach for treating substance abusers. Gardner Press, New York, N.Y.

Yalom, I.D. (1975). The theory and practice of group psychotherapy. Basic Books, New York, N.Y.

Yohai, S.J. and C. Winick (1986). AREBA-Casriel Institute: A third-generation therapeutic community. J. of Psychoactive Drugs, 18(3), pp. 231-237.

Zimmer-Höfler, D. and A. Widmer (1981). Democratically or hierarchically structured therapeutic community for heroin addicts. In: The therapeutic community in various cultures worldwide. Proceedings of the 6th World Conference of Therapeutic Communities, pp. 92-98, DARE Foundation, Astrad Production House, Manila.

Zimmer-Höfler, D. (1987). Systemische Gesichtspunkte im therapeutischen Umgang mit Drogenabhängingen. Buchbeitrag für Reiter und Brunner Hrsq. Springer (1988). Serie A nr. 17 Wissensch. Inform. März 1987. Forschungsgruppe des Sozialpsychiatrischen Dienstes, Zürich.

Zinberg, N.E. (1984). Drug, set and setting: the basis for controlled intoxicant use. Yale University Press, New Haven.

Zuckerman, M. (1970). Drug usage as one manifestation of a sensation seeking trait. In: Keup, W. (ed.) Drug abuse: current concepts in research. Charles C. Thomas, Springfield, Ill.

Zuckerman, M. (1986). Sensation seeking and the endogenous deficit theory of drug abuse. In: Szara (ed.) Neurology of Behavioral Control in Drug Abuse. NIDA Research, Monograph Serie 74, pp. 59-70. U.S. Gvt. Printing Office, Washington D.C.

Zuckerman, M., E.A. Kollin, L. Price and I. Zoob (1964). Development of a sensation seeking scale. J. of Consulting Psychology, 28, pp. 477-482.

Zuckerman, M., Sola, S., Masterson, J. and J.V. Angelone (1975). MMPI patterns in drug abusers before and after treatment in therapeutic communities. J. Consult. Clin. Psychology, 43, pp. 286-296.